The Michigan Historical Reprint Series

Reprints from the collection of the University of Michigan University Library

This volume is produced from digital images created through the University of Michigan University Library's preservation reformatting program. The Library seeks to preserve the intellectual content of items in a manner that facilitates and promotes a variety of uses. The digital reformatting process results in an electronic version of the text that can both be accessed online and used to create new print copies. This book and thousands of others can be found in the digital collections of the University of Michigan Library. The University Library also understands and values the utility of print, and makes reprints available through its Scholarly Publishing Office.

For access to the University of Michigan Library's digital collections, please see http://www.lib.umich.edu.

The Scholarly Publishing Office seeks to disseminate high-quality, cost-effective scholarly content through both print and electronic publishing. Information about the Scholarly Publishing Office can be found at http://spo.umdl.umich.edu.

The Scholarly Publishing Office
the University of Michigan
University Library

J.D.FELTER·SC

THE

AMERICAN

BEE KEEPER'S MANUAL;

BEING A

PRACTICAL TREATISE

ON THE

HISTORY AND DOMESTIC ECONOMY OF THE HONEY-BEE EMBRACING A FULL ILLUSTRATION OF THE WHOLE SUBJECT, WITH THE MOST APPROVED METHODS OF MANAGING THIS INSECT THROUGH EVERY BRANCH OF ITS CULTURE, THE RESULT OF MANY YEARS' EXPE-RIENCE.

BY T. B. MINER.

EMBELLISHED BY THIRTY-FIVE FINE WOOD ENGRAVINGS.

Queen.

Worker.

Drone.

SECOND EDITION.

NEW YORK:

PUBLISHED BY C. M. SAXTON,

121 FULTON STREET.

1850.

C. W. BENEDICT, *Stereotyper,*
201 *William street, cor. of Frankfort.*

PREFACE.

The following treatise has been written to fill a vacuum in this country, that has long existed. How it has happened that the management of the honey-bee should have been so neglected by writers in the United States, I am at a loss to comprehend; but so it is, and we cannot boast, up to the present time, a single volume on this subject, worthy of being called a full, practical treatise on the culture of this insect.

Such small essays as have appeared from the pens of American authors, have given little, or no information of practical utility; the most of them, not even presenting a single engraving, as an illustration. In this work, the expense of the embellishments alone, will equal the *entire* cost of publishing any of the works of American origin, that have preceded it; and it has been my pride and aim, to render it a production, that will not only compare with, but even exceed the most popular European treatises on the same subject, especially in all matters of a *practical* nature.

The great difficulty in the way of producing a truly popular work on the honey-bee, has hitherto been, the imaginary *dryness* of the subject, operating as a great discouragement to practical apiarians to write thereon;

but in this work I have, I think, divested the subject of its dryness, to some extent, and I have placed it before the reader in a more attractive form, I trust, than it has hitherto appeared in many works of this kind. I have endeavored to discuss the various questions in a clear, ample, and comprehensive manner, divested of the superstition of the ignorant, and the errors of those who profess to be learned in the science.

I have not placed that implicit confidence in all of Huber's writings, as may be seen, that some apiarians do; yet I have quoted him, to some extent, on subjects where he is undoubtedly correct.

What I have written in the following pages, is mostly the result of my own practical experience, during many years of close application to the management, and the study of the honey-bee. On some points I have taken an entirely new course, in my own management of bees. For instance, I treat them differently in the winter season especially, from the ordinary custom, keeping them much cooler, &c.; and my general course of management is peculiar to myself, with a full knowledge of all systems, yet based upon the true principles of the nature and economy of the bee. My success in the culture of this insect, has been beyond precedent, and having the test of a long series of years to support me, I offer this work as one worthy of the confidence of the public.

THE AUTHOR.

CHAPTER I.

THE QUEEN.

Every association of bees is composed of three classes, viz: a *queen, drones* and *workers*, and when separated from their natural connections, they loose all their attributes of industry and soon perish in inaction.

The queen is the mother of the entire increase of every family of bees, unless in rare cases of the existance of a few fertile workers, that produce drone eggs only, of which I shall speak hereafter. The queen is longer than either drones or workers, and much larger in every respect than a worker, but not so large as a drone. Her trunk or body is shorter than that of the other two classes, and her abdomen tapers to a point, in the form of a sugar-loaf. Her legs are longer than those of drones and workers, but have no cavities or baskets for holding gathered stores. The most remarkable feature pertaining to her majesty, is the shortness of her wings, reaching only to about two-thirds of the length of her abdomen. Her color is much darker than workers, and sometimes approaching to a jet black; that is, upon her upper surface, but her belly is of a dark orange color. It is this latter hue that enables one to easily distinguish her in a cluster, even without seeing any other part of her body.

WHEN SEEN, AND HOW FOUND.

It is only in particular instances that the queen is to be seen, such as during swarming, or on her ærial excursions, which takes place on the second or third day after being hived, or upon some occasion of her being found in a cluster of bees upon the alighting board. In this latter case, which occurs with recent swarms only, as a general rule, and very seldom, a close cluster of bees is seen about the size of a hen's egg, remaining quiet, and when the feather end of a quill, or a stick is used to separate them, and they instantly re-form into a cluster again, it is almost certain that the queen is in the centre. The kind of cluster that I allude to, is very different from ordinary clustering upon the side of the hive, or on the bottom board, when the bees are driven out by heat; then the bees cluster with their heads upward; but in clusters where the queen is to be found, nothing of such a regularity is to be seen.

STING OF THE QUEEN.

The queen is armed with a sting which is curved, but she seldom uses it, except against rival queens. Indeed, she may be taken with the bare fingers, at any time, with perfect impunity; but a worker taken in that manner, would be dropped as a piece of hot iron.

HER FECUNDATION.

The fecundation of the queen has ever been a subject of deep interest to naturalists, and it is not at the present day so fully decided, in what manner, or by what agency it is effected, as to put the question entirely to rest; and I may safely add, that the day will never come, when this long disputed point will be so fully cleared up as to silence all opposition to the now generally received opinion of the case.

Some naturalists and apiarians have supposed that the queen is *self*-impregnated; that is, that the fecundating germ of the ovary is inherent in her, and when her eggs are laid, that the drones fertilize them, and generate the principle of animal life by incubation, or sitting upon them. Others have supposed that a vivifying *seminal aura* exhaled from the drone, penetrates the body of the queen, and that produces impregnation. This opinion arose from the fact, that a strong odor is sometimes exhaled from them.

Naturalists rightly supposed, that a sexual union did take place between the queen and drones in some manner, but how, or when, was beyond the scope of their knowledge, since such an union had never been beheld by mortal eyes. However, during the latter part of the eighteenth century, light seemed to dawn upon this long hidden mystery, which had lain shrouded in darkness for thousands of years. The fact that the sexual union of many species of winged insects takes place in the air, while on the wing, did at last, after centuries upon centuries had past in wild speculation, cause the films to fall from the eyes of the naturalists of the day, and they came to the conclusion that the impregnation of the queen bee must be effected in the like manner.

That many hundred years should have past away, before this simple fact should have become developed, is truly a matter of surprise! Yet it is a matter of still greater surprise, that many apiarians of the present day pertinaciously adhere to ancient notions in regard to the agency of the drones in the impregnation of the queen, and utterly refuse to divest themselves of traditions founded in error and superstition.

That such is the natural use and purpose for which drones were created, viz: to effect impregnation on the wing, I presume the reader will readily believe, on hearing what I have to say on the subject hereafter, in Chap. iii. devoted to "*drones.*"

HUBER, THE GREAT (?) APIARIAN.

Huber, a German naturalist of distinction, who flourished at the close of the eighteenth century, has placed this question beyond a doubt; provided that we may place confidence in his statements, which I consider somewhat questionable.

Since Huber is cited as orthodox authority by almost every writer on the honey bee, or at least by a very large portion of them, and inasmuch as many eminent naturalists and apiarians consider the greater portion of his writings as an ingenious fabrication of experiments, that never occurred save in the imagination of this naturalist, or of his assistant, I deem it necessary to place the position of this author fairly before my readers, that they may be able to judge for themselves, in some measure, whether he is, or is not, entitled to full credence. I do this, partly for the reason that some authors on this subject, within the reach of an American public, servilely tread in the footsteps of Huber, without ever having read his writings from his own pen, which is perfectly apparent, from the limited knowledge of his work, possessed by the writers, as their essays plainly manifest.

At the time that Huber wrote, about the year 1790, the natural history of the honey bee, as well as its domestic economy and management, was in a state of obscurity. Very few men of talent had given the subject a profound attention, and the traditions and absurd fancies of olden times, in regard to this insect, were believed and acted upon, by the majority of bee-keepers. At this epoch, Huber professed to have made a series of experiments, during a period of some five or six years, illustrating the physiology and economy of the honey bee to an extent that had never been reached before. But his writings throw no light whatever upon the domestic *management* of bees; therefore, they are of no value to the apiarian who studies the economy of bees, merely for the *profit* derived from them. The naturalist alone considered his discoveries as highly important and valuable, and being a novelty, the world at once took the truth of his theories and experiments for granted, and Huber was forthwith placed upon the pinnacle of apiarian science.

Many apiarians who subsequently wrote upon the bee, servilely followed him through both truth and fallacy, without being able, from their own experience, to either refute or corroborate his theories and hypotheses. Encyclopœdias and other publications cited him as unexceptionable authority, and he was styled the "Prince of Apiarians;" hence we find American authors taking their cue from some foreign proselyte to his theories, and blindly re-echoing many of his discoveries as facts, which may be as far from the truth, as the east is from the west!

The reader may here inquire, if the natural history and domestic economy of the honey bee, is so involved in mystery and obscurity, as not to be fully understood at this late day, and susceptible of being clearly expounded and laid down, without the possibility of error? Yes sir, it is thus involved; and the day will never come, when the veil of obscurity that now shrouds much pertaining to this interesting little insect will be wholly removed.

Man may experiment—he may send forth theory and hypothesis to the end of time; yet the natural instinct and wisdom of the bee, in many of her acts, and the *modus operandi* of her internal domestic labors, to a great extent, will forever be *terra incognita* to all human knowledge!

Let not the reader suppose from the above remarks, that we are doomed to remain ignorant of important facts, to enable us to meet with perfect success in our management of bees—the curtain *has* been raised, and man has beheld—*enough for man to know.*

As the wisdom of God is past finding out, so is the instinctive wisdom of the little bee, a direct attribute of the Architect and Creator of all animate and inanimate nature, beyond the pale of human knowledge.

HUBER'S AUTHORITY DOUBTED.

As it will be necessary for me in the following work, to frequently allude to Huber and his writings, since the history of the bee is based, to a great extent, upon the foundation laid by him, the reader will excuse a continuation of remarks touching the confidence due to his statements. His writings comprise simply a series of *letters* to his friend and patron, Bonnet, of Geneva. Bonnet's reputation as a naturalist stands high, and those letters were written at his suggestion of various things pertaining to bees, then in obscurity, and which, for the benefit of science, it was necessary to unfold. Huber being in affluent circumstances, and unable to attend to any ordinary pursuits, in consequence of his blindness, he being unable to discover the difference between a white person and a colored one, he, with the aid of a servant, instituted his experiments in the economy of bees, to avoid that *tedium vitæ* that ever accompanies the unemployed.

Now, had Huber had personal ocular demonstration of what he has written, as being verified by him, through his assistant, we might consider him entitled to credence; but he trusted entirely to his servant, in all those alleged discoveries that have astonished and amazed the world.

I can give but a faint credence to discoveries thus verified, so far as the *authority* is concerned; but where Huber's statements tally with well known principles, we should give him the benefit of our confidence in his assertions.

The reader may be interested to know what wonderful discoveries this man has made? They consist in discoveries relative to the impregnation of the queen,—retarded impregnation and its effects—verification of the existence of fertile workers—the power of the bees in raising a queen from any ordinary worker's egg at pleasure—combats of rival queens—massacre of drones, &c., &c., interwoven, as many apiarians presume, with considerable fiction, since many things which he alleges to have seen, or rather that his *servant* saw, have never been beheld by any one else.

HUISH'S OPINION OF HUBER.

Huish, a writer of some celebrity on bees, whose work was published in London, in 1844, says, "Huber, from a natural infirmity of the eyes, was wholly disabled from prosecuting his researches into the natural economy of the bee, and consequently that he relied solely on the skill and information of his servant, François Beurnen's, for the veracity of those singular discoveries, which, under the sanction of his name, have been sent forth into the world, but which will never stand the test of a rigid and scientific examination.

Now, this same François Beurnens was a rude, uneducated Swiss peasant, with a mind immersed in all the prejudices of his country, and who pertinaciously adhered to many of the Swiss customs in the management of bees, which have for their basis the grossest ignorance and superstition. Thus, for instance, when any of the family died in which Beurnens was a domestic, he turned all the hives in the garden topsy turvy, in which condition they were obliged to remain until after the funeral, as it was most proper and becoming that the bees should be made to sympathize with the loss which the family had sustained."

Notwithstanding that the lash of ridicule has been well applied to Huber, by those apiarians whose experience has proved a portion, at least, of his writings as fallacious, yet some of his discoveries are undoubtedly true, inasmuch as they accord with the observations of apiarians in general; and of this kind is the discovery of the manner in which the impregnation of the queen takes place in the air, by the drones, the subject on which I was speaking, that gave rise to the introduction of this author; and I think I cannot more profitably occupy the attention of the curious reader for a few moments, than to give his account of this discovery in his own words. Here it is:—

HUBER'S DISCOVERY OF THE IMPREGNATION OF A QUEEN.

"Aware that the males usually leave the hive in the warmest part of the day, in summer, it was natural to suppose that if the queens were obliged to go out for fecundation, instinct would induce them to do so at the same time as the others.

"At eleven in the forenoon, we placed ourselves (*Beurnens was the one to watch for the queen, directed by Huber, the reader will understand; yet Huber always wrote as if he could see,*) opposite to a hive containing an unimpregnated queen, five days old. The sun had shone from his rising, the air was very warm, and the males began to leave the hives. We then enlarged the entrance (*Huber had contracted the entrances of several hives to prevent the egress of the queens,*) of that selected for observation, and paid great attention to the bees entering and departing. The males appeared and immediately took flight. Soon afterwards the young queen came to the entrance; at first she did not, but during a little time traversed the board, brushing her belly with her hind legs, neither workers nor males bestowing any notice on her. At last she took flight. When several feet from the hive she returned and approached it, as if to examine the place of her departure, perhaps judging this precaution necessary to recognize it; she then flew away, describing horizontal circles twelve or fifteen feet above the earth. We contracted the entrance of the hive that she might not return unobserved, and placing ourselves in the centre of the circles described in her flight, the more easily to follow her, and witness all her motions. But she did not remain long in a situation favorable for our observations, and rapidly rose out of sight. We resumed our place before the hive; and in seven minutes the young queen returned to the entrance of a habitation which she had left for the first time. Having found no external evidence of fecundation, we allowed her to enter. In a quarter of an hour she reappeared, and after brushing herself as before, took flight, then returning to examine the hive, she rose so high that we soon lost sight of her. This second absence was much longer than the first, it occupied twenty-seven minutes. We now found her in a state very different from that in which she was after the former excursion; the organs distended by a substance, thick and hard, very much resembling the matter in the vessels of males, completely similar to it in color and consistence."

Huber afterwards says, that from subsequent discoveries, he found that what he took for the generative matter, was the male organs left in the body of the female.

QUEEN'S FLIGHT TO MEET THE MALES.

That queens do thus sally forth on the second or third day after entering a new habitation with a swarm, is a fact that has come under the observation of many apiarians, yet it is doubtful whether the change in the appearance of them on their return, as spoken of by Huber is generally, if ever visible. If the young queens are to be seen at all, it is at this period, and it is not unfrequent, that queens of all swarms, *after the first,* during the first few days of their inhabiting their new tenement, are found in a cluster of bees at the entrance of the hive or near it. The reason of this is, that on the return of the queens from their excursions in search of drones, they are immediately surrounded by their subjects and held prisoners for a brief period. The reason why the queens of *first* swarms are not thus found is, that such swarms are accompanied by *old* queens, whose impregnation is already effected.

ANALOGY PROVES IMPREGNATION ON THE WING.

It is a well known fact that the sexual union of the humble bee takes place on the wing. I have frequently witnessed it; and it is the same with the most of insects of the winged tribe; hence analogy is strongly in favor of the theory of the impregnation of the queen honey bee as aforesaid.

HUBER CONFINES QUEENS TO PROVE THE THEORY OF IMPREGNATION ON THE WING.

Huber states that he confined the queen with a large number of males, and also confined her with the males excluded from the hives; at the same time admitting the ingress and egress of the workers as usual, and in every case, which were numerous, the queens remained *sterile.* He confined them over a month, which was enough to test the question whether a queen can be fertile and not leave the hive.

RETARDED IMPREGNATION.

Huber also states, that when a queen is retarded *twenty-one* days from her birth in her impregnation, she then, and ever thereafter, lays drone eggs only.

As no one has ever experimented on queens in the manner of the above two cases,—at least, no one having yet given publicity to any observation, refuting or corroborating Huber's discovery, it is not easy to say, whether the last case be true or false.

In regard to the sterility of queens that have not been allowed to leave their hives, there is no doubt. Their impregnation being effected on the wing, it follows, of course, that confinement with, or without males, must render them barren.

That retarded impregnation does cause queens to lay drone eggs, is quite possible, yet the fact might not come under the observation of an ordinary bee-keeper in a century, in most cases, since nature has so amply provided for the effectual impregnation of queens, that to be retarded by any natural event, is out of the question, except in cases of the death of a queen, at a period when the drones are exterminated, or so few of them existing as to jeopardize the impregnation of the successor to royalty.

I, myself, have had a case in which drones only were produced, but whether it was owing to a retarded impregnation of the queen, or whether it was the production of fertile workers, I am unable to say positively, owing, unfortunately, to the destruction of the stock, by my own act, before the question could be decided.

I shall give the full details of this singular case in my remarks on "*workers,*" as it more properly belongs to that class of bees.

CHAPTER II.

THE WORKERS.

THE workers are the smallest bees of the family. A worker's head is of a triangular shape, as well as that of the other classes,—the abdomen is connected with the trunk or thorax, by a small ligament, of a thread-like nature, and it is composed of six scaly rings, at the apex of which, is the sting, which is full of barbed points like an arrow, which can only be seen by the aid of a strong magnifier, and which prevents the extraction of the sting when darted into one's flesh, causing a portion of the entrails of the bee to be drawn out with it, and thus causing death to the insect.

Every bee has *four* wings; and on queens the number of wings is more preceptible than on workers or drones. They have six feet.—The eyes are situated upon the upper surface of the head.—Every bee has a pair of *antennæ*, of a fine wiry flexible nature, protruding diagonally from the head, which are used as organs of *feeling*, or perhaps of *smell*, since a *stranger-bee* is known at once, on applying the antennæ to it.

The antennæ of the queen generally are turned or curved *downward.*—This is their natural position, and the inexperienced bee-keeper may know her majesty from this circumstance, when he is in doubt as to her identity.

Workers have spoon-like cavities or baskets upon their posterior legs, that hold the pollen or farina gathered by them. No other bee has these cavities. Workers also have a honey bag, or stomach, expressly to hold the gatherings of the day. It will hold about half a *drop* of honey. The bodies of bees are covered with a hairy down, which, through a microscope, appears like a defence of *palisades.*

Wonderful are the labors of this class, and truly may they be called "workers," for never did industry show a brighter example of indefatigable perseverance, than in the labors of this little insect.

The following little stanza often recurs to one's mind as he surveys these ever industrious workers, hurrying to and fro, on a bright sunny day.

"How doth the little busy bee,
  Improve each shining hour;
Gathering honey all the day,
  From every opening flower."

The workers are the architects of the association. They construct the cells, arrange their size and distances, repair damages, &c., &c. They are the laborers of the family; they gather the honey and farina, and compound the food for the young bees, and upon their skill and labors depend the prosperity of the colony.

Who that has witnessed this class of bees, during the height of their harvest, has not been forcibly impressed with their indefatigable industry! They sally forth before the rising of the sun, and return when evening twilight has cast her sombre mantle over the face of nature, laden with sweets, which but for this industrious insect, would be lost on the desert air. Neither the scorching rays of a vertical sun, nor the peltings of the storm, can restrain their zeal in securing to themselves life and prosperity, by availing themselves of every moment that can possibly be employed, when the fields are decked with the flowers that most invite them.

They do, indeed, afford a theme worthy the attention of the philosopher and moralist. Man is here taught a lesson that should never be forgotten; but ever be indelibly impressed on his mind. The improvident and lazy may here learn, from the book of nature, truths that would lead them to fortune and prosperity, were not their consciences seared and callous to all lessons of wisdom.

The little bee, aware that the days of her harvest are few, "makes hay while the sun shines," and that Divine injunction, "Whatsoever thy hands find to do, do with all thy might," is here acted upon, and carried out to the letter, to the shame of man, for whose especial benefit it was given.

To the bee, no written law can be given by their Creator; consequently, an *instinct* is given them to guide them in their labors; and when the flowers are faded and gone, and the bleak blasts of winter flit around, she looks upon her loaded combs, as the reward of her toils, and laughs at the raging winds and pitiless storms.

But how stands the case with man—the being who is made but a grade inferior to Angels? Does *he* show himself worthy of his vocation—does he even show himself equal to the little puny honey bee, in foresight of those evils that delay, neglect, procrastination, inaction, or downright *laziness* produce?

For an answer, just cast your eye around.—In yonder hovel is a human being clothed in rags, surrounded by a large family of children, who are crying for bread. The emaciated mother, the unwilling victim of the father's improvidence, is fast approaching the grave. Her leaky tenement has, year after year, caused the seeds of disease to germinate, and now friends call to console—to alleviate; it is too late. Ah! how is this? has this man had his health—has he had the use of his limbs, in this land of prosperity, where poverty need be known only in name, to be thus impoverished, and to have his house falling around his head? Indeed, he has been as hale and hearty as the most robust among us. He is also an excellent workman, but he has never heeded the old adage, "make hay while the sun shines;" and when winter comes, it finds him naked and penniless—his children cold and hungry, and his wife without the ordinary comforts of life. Would he but follow the example of the little bee, and from her learn wisdom, poverty would be banished from his door, and the bleak winds of winter would bring no terrors, and their howl would be music in the ears of the little fire-side group, as they sing their merry songs of contentment and happiness.

THE EFFECTS OF A SUDDEN STORM ON BEES.

I have often seen these workers returning so late in the evening, in warm sultry weather, that they were barely able to find their respective hives; and so eager are they to devote every moment to their labors, that many of them, suffer themselves to be overtaken by the tempest and storm, before they take their homeward flight.

It may be supposed, that under such circumstances, storms and winds arise so suddenly, that the bees are taken by them unawares; but such is not the case.

Wishing to note particularly the return of bees from the fields, in the height of their harvest, and to what extent they would remain out, on the approach of a heavy thunder storm, I, in the month of June last, took a station among my hives, on the approach of a shower, and minutely watched their course. It was about the middle of the day, or noon; the sun had been shining all the morning, and the bees were out in their greatest numbers.

On the appearance of dark clouds, in the west, and accompanied with thunder, the bees commenced returning more than is usual in fair weather. In about a half an hour, the heavens were darkened by clouds, with a slight sprinkling of rain, and the roar of thunder shook the earth. At this crisis, the bees came in with a rush and a few, in the face of the approaching storm, darted forth to the fields again.

This state of things lasted forty minutes, with sufficient rain to have given every bee full warning, even were they both blind and deaf.

Even the most distant bees, I considered within the reach of the rain, and I supposed, that in fifteen minutes from the commencement of the shower, every bee would have been in; but such was not the fact. They continued to pour in during the whole of the forty minutes; then the winds commenced blowing furiously, and the rain fell fast; I took an umbrella, and standing in the midst of the apiary, beheld the bees beating in against winds and rain, until the water came in such torrents, that a perfect sheet encompassed me; and at this juncture, several bees on their return, finding it impossible to gain their hives, came under my umbrella for protection. Every bee that was out at that crisis, must have been dashed to the ground, unless they sought refuge on the nearest thing that came in their way.

This observation proved that bees can fly a considerable distance to their homes, while the rain literally pours down. Before the last heavy dash to which I refer above, I noticed the bees coming in very slowly indeed, for the rain came down in torrents; yet they did slowly make headway through it. Their speed, as they approached the apiary, was much slower than a man usually walks; and I presume, that it would have been impossible for them to have proceeded much farther.

This observation also shows how indefatigable they

are, in the pursuit of their natural avocation. The sturdy iron-bound frames of the laborers of the adjacent field had taken flight, long before the bees considered it necessary to vacate the flowery hills and vales, as if those iron frames were made of salt, while the little frail bee, with her fragile silken wings, braved the tempest, and bid defiance to the driving storm!

THE SEX OF WORKERS.

Much diversity of opinion has been expressed, in regard to the *sex* of workers, by naturalists and apiarians; and this is not the only question in dispute among them. The natural history, physiology, and economy of the honey bee, has perplexed and baffled more scientific men in their attempts to unveil the secrets of their nature, than any other subject whatever. As I before stated, much that pertains to the bee, is beyond the pale of man's knowledge; and a thousand years hence, darkness and mystery will hang over this subject, and man will behold and wonder;—but to fathom the secrets of their intuitive wisdom, he never will be able.

The reader may possibly ask, "what benefit is it to know, whether the workers are *males*, *females* or *neuters*, so long as we know sufficient to enable us to manage our bees with perfect success?"

Why, sir, so far as *pecuniary* advantage is concerned, it is of no consequence to know many things concerning the bee, that will occupy much of my attention in these pages; but there is a curiosity extant, that is not satisfied with any thing short of all the knowledge, touching the nature and habits of this insect, that is attainable by man; and while many will pass these pages, with a hurried glance, for those that reveal a knowledge, that comes home to the *pocket* of the reader; saying, "why is this long useless expenditure of words upon *queens, workers, drones, fecundation, sex of workers, &c., &c.,*" others will wish for a more lengthy and elaborate treatise, on the same subject.

The sex of workers is neither *male* nor *female*. They appear to be strictly a phenomenon in nature, and by many, are termed *neuters*.

WORKERS SAID TO BE SOMETIMES FERTILE.

The workers approximate very nearly in their internal organization, to the queens, having ovaries like them, but not so fully developed. In their natural capacity, they never produce eggs; yet it is contended, that under peculiar circumstances workers exist, partaking of the nature of queens, to a much greater extent, than in their ordinary state; and that such workers lay *drone* eggs only. The most positive proof of this assertion, ever given to the public, so far as I have been able to learn, is adduced by Huber. He states, that having a hive in which drone eggs only were produced, and believing its legitimate queen to be lost, his servant caught every bee in the hive, examined them carefully, made them show their stings, in order to test their gender, as *small* males are sometimes found that very nearly resemble workers, which males have no sting; and he then put them into a glass cylinder; and so on, to the very last, and not a bee was found, except workers. Indeed, he experimented on two hives in this way, as he says; and so tedious was the job, of catching and examining the bees, that it took *thirteen* days to perform the operation.

From this experiment, he says, that he was certain that workers do sometimes produce drone eggs, as before stated—in short, his servant, Beurnens, actually took one in the very act of laying.

How far we can credit Huber's statements in regard to this transaction, I cannot say. His hives were of the kind termed *leaf hives*, which he was enabled to open, like the folds of a book; and it is possible, that the operation of catching the bees, may have been performed; but I doubt whether it will ever be done again.

FERTILE WORKERS, AND THE POWER OF WORKERS TO PRODUCE QUEENS FROM ORDINARY WORKER EGGS!

The manner and cause of the production of workers that lay drone eggs, is as follows:—provided that such do ever exist, though I came within an ace of verifying the fact myself, as I shall relate.

It is necessary here to inform the reader of the power of workers in forming, or producing a queen, in order that he may rightly understand the question.

KINDS OF EGGS LAID BY THE QUEEN.

The queen lays but two kinds of eggs, viz: drone and worker eggs; and when queens are wanted, ordinary worker eggs are laid by the queen, in cells made expressly for royal use, termed *queen cells*. Here is a cut of a royal cell, precisely as taken from one of my hives:

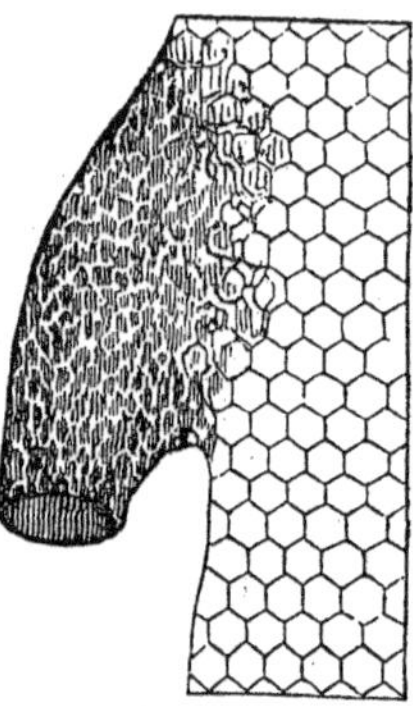

ROYAL CELLS—HOW CONSTRUCTED.

The queen cell is of the exact natural size and shape as it here appears; but the worker cells are on rather too small a scale to give the *tout ensemble* in its regular proportions, but that is of little account, since the only object of the cut is, to illustrate the position and natural shape of royal cells alone.

It will be perceived, that this cell hangs vertically; that is, with the mouth downward. These cells are generally built upon the edges of the combs; and for this purpose, the bees leave one side of their combs, without much support along the edges, except an occasional bar, or brace, while the opposite edges are firmly cemented to the hive, through their whole length. The distance between the combs that are intended for royal cells, and the side of the hive, is from a quarter to a half an inch; giving just room enough for these cells, which are about the size of a peanut, and look, in shape and outward appearance, very much like this nut, with one end cut off, and the nut extracted. The bases of these cells, however, are broader than a peanut; and the shape is somewhat like a sugar loaf, placed upon its small end. Royal cells are also constructed in the centres of combs, on the edges of passages through them. They who have been in the habit of cutting out combs, have undoubtedly perceived small orifices, about the size of a half dollar, through which the bees pass from one comb to another, and on the edges of these apertures, royal cells are as frequently built, as on the edges of the combs. Why such a large, cumbrous cell is necessary for the raising of queens, that are of less size than a drone, is very singular! There is more material put into one of these royal tenements, than would be required to build a dozen drone cells; and then, they must hang *downward!* Here is one of the mysteries pertaining to bees, that man can never unfold. A drone cell, lengthened a little, would be just the thing for these young queens; yet these stubborn bees will not be taught *improvement;* they seem so attached to the customs of their forefathers.

HOW YOUNG QUEENS ARE PRODUCED—THE NUMBER OF ROYAL CELLS IN A HIVE, ETC.

When the young queens are wanted, several of these cells are constructed; say from five to twenty-five; and the queen deposits worker eggs therein at intervals, so as to mature, at about the period that they will be wanted to go off with swarms. I have noticed some difference of opinion in regard to the largest number of royal cells, ever found in a single hive. Huber is denounced by Huish, for asserting that he discovered *twenty-seven* in a single hive; and he (Huish) contends, that no hive ever contained at once, more than *seven*; but I have myself, the present season, taken *twenty-two* from one hive, and *seventeen* from another, but they were not all perfect cells. There are always more or less royal cells that are not completed, in every case; for, as soon as a certain number are so far advanced, that the young sovereigns are sure of being perfectly developed, the remaining cells are discontinued.

It should be borne in mind, that these royal cells are not completed, and then made the receptacles of the eggs; but when about half constructed, they receive the egg; and as the larvæ* progress, the cells are completed. These half-constructed cells, resemble an acorn, devoid of the nut.

DIFFERENT FOOD PROVIDED FOR THE YOUNG QUEENS.

When the royal cells have received the eggs, and they become hatched out, the workers provide different food for the larvæ from that which is fed to workers and drones. This food, which has been termed *royal*

**Larvæ* is the term given to the worms or grubs until the cells are sealed. From the sealing of the cells, to full development, a nymph, pupa or *crysalis;* yet the term *larvæ* is properly applied, by some naturalists, during the whole period of the embryo state.

jelly, immediately changes the nature of its recipient, and the properties of a queen begin to be unfolded. The size of the cell, and its vertical position, perhaps has some influence and agency, in producing a royal scion; but the grand elixer, is the *royal jelly,* as is universally supposed, that effects the change.

The only man that ever pretended to have actually discovered this royal pap, is Huber; for there was nothing under heaven that he could *not* discover, through his assistant, Beurnens, who used "*to upset the bee hives on the death of a member of the family!*" He not only *saw* it, but *tasted* it, as he states.

Huber gave it the name of *royal jelly,* and on giving his discoveries to the world, naturalists promulgated his theory, and many apiarians have become re-echoers of it. Some, indeed, do contend, that it takes a *queen egg* to produce a *queen,* but they are behind the age. This assumption will not stand a moment. I have time and again, proved that a *worker egg will produce a queen!* as I shall relate, at the proper time. And in regard to the theory of "*royal jelly,*" it is quite plausible, for, if it be not a different food, that produces queens, *what can it be?* No man, in my opinion, knows anything more about this royal jelly, or whether it does or does not exist, than perhaps the reader, who perchance does not know a *queen* bee from a *worker* or *drone*; yet the great and wonderful truth stands impregnable, that a different treatment *does* produce queens, and that *positively,* there is no difference in the *egg* used for this purpose, and that, from which a worker emerges!

We *must*, then, come to the conclusion, *that it is the food that makes the change;* and we will continue to affirm, that it is the royal jelly, that effects the change, not at all fearing, that any one will ever be able to prove us to be in error, as it is not in the power of man, ever to go beyond simple *conjecture* on this point.

THE FORMATION OF A NEW QUEEN IN THE PLACE OF ONE THAT DIES, OR IS LOST.

The bees having the power to make queens at will, from worker eggs, it follows as a natural consequence, that in the case of the death of a queen, or of her loss when absent from the hive, which does sometimes happen, *they can at once supply her place,* provided that she left any eggs, or larvæ *less than four days old.* Here we find one of the most wonderful provisions of nature, pertaining to the natural economy of the bee; for, were they not able to thus replace the loss of a queen, this insect would soon become extinct.

There are seasons, however, in which the queen may die and leave no eggs, or larvæ behind her under four days old; and in such cases, the family must perish, unless supplied with a new queen by their proprietor, or a piece of comb, containing eggs, or larvæ of a suitable age; and in such a case, the proffered comb, if properly attached in the hive, in a natural position, answers every purpose of larvæ left by the queen. But such seasons or instances are not frequent with well peopled hives, for larvæ may be found in such hives, to a greater or less extent, almost every month in the year. Even in the dead of winter, larvæ have frequently been found in the centres of very strong stocks* or swarms; and it appears to be thus ordained by nature, in erder to always admit of the bees being able to provide against the loss of their sovereign. In their natural state in the forest with an abundance of room, perhaps they never experience the loss of a queen, without being able to replace her, except in cases of small swarms issuing, in which case, they would be liable to the same casualties of domestic swarms, until they have existed a season or two and have become numerous.

THE SUPPOSED CAUSE OF THE FORMATION OF FERTILE WORKERS.

The reader now having a little insight into the manner in which queens are made, I will proceed to state in what manner these semi-fertile workers are supposed to be produced; for, I must inform the reader, that all the insight that has ever yet been obtained on this subject, is nothing more than *simple conjecture and hypothesis.* This is, as I have observed, "*terra incognita,*" or unknown land, to the apiarian explorer, and may be set down as one of the unfathomable mysteries of the nature of the bee.

The royal cells being constructed, or in progress of construction, and containing the larvæ to be transformed into queens, and being fed on the royal jelly, as afore-

* Every family of bees is termed a *stock*, after the first year of their existence, and a *swarm* during the first year or season.

said, it is supposed that on some occasions, that the worker larvæ, situated immediately adjoining the royal cells, may, either by accident, or otherwise, be fed a little of the royal pap, which, not being sufficient to produce queens, and only enough to so far develop their ovaries, as to enable them to lay *drone*-eggs only.

It is not probable, even if the above hypothesis be true, that workers would become sufficiently fertile to be able to lay both drone and worker-eggs, by being wholly fed on royal food, since the shape and position of a royal cell, has its peculiar effect upon its tenant, otherwise such cells would not be constructed, as bees do nothing without a good reason.

A CASE OF RETARDED IMPREGNATION IN THE QUEEN, OR OF FERTILE WORKERS, COMING UNDER THE AUTHOR'S OWN OBSERVATION.

I will now relate what took place under my own immediate observation, iu regard to the laying of drone-eggs in one of my hives.

On examining one of my hives early the present season, (1848) I found a swarm of last year in a very weak condition, not having above two or three hundred bees in it. How this diminution in numbers happened, or what the cause was, I could not imagine; since the swarm was large, and in good condition apparently, last fall; having filled the hive with comb, and having laid in an abundance of honey for winter consumption. I closely watched this hive, to ascertain whether any of the few bees it contained gathered farina; as that fact would throw some light on their condition, since where the queen is lost, the bees never gather this food of the larvæ, because they have no necessity for it, while in that condition. I did perceive an occasional bee enter with pellets of farina, and I at once took it for granted, that the queen was among them, and that she would prove fertile; but owing to the very small number of bees composing her family, I was aware that it would be very late in the season, before she would be able to replenish the hive in numbers, owing to the difficulty in generating the necessary animal heat. After watching during the month of May, in vain, for any apparent increase, I concluded that if the hive remained much longer in that condition, the moths would take possession, and give the handful of bees therein "notice to quit;" and if they should manifest any disposition to refuse to comply with so reasonable a requisition, a "writ of ejectment" would speedily follow; and not wishing to have any controversies arise between my bees and so stubborn a creature as the moth, touching the right of possession, I immediately commenced cutting out a portion of the combs, in order to give the bees a better chance to defend themselves, in case of being intruded upon.

In cutting out these combs, I discovered in one of the centre combs, near to the top of the hive, a piece of brood, about two or three inches square, which was entirely *drone*-brood. I searched in vain for any trace of worker-brood, nor did I find a solitary worker larva, up to about the 20th of June, when the family was destroyed; but I found a small increase of drone-larvæ, and the most of what I originally discovered, regularly matured.

On making the discovery of drone-brood, I searched in vain for the queen, and being able, with the feather end of a quill, to almost bring every bee in sight, and after many attempts at her discovery, not seeing any signs of royalty, save the brood as before stated, I came to the conclusion, that I had a veritable instance of the fecundity of workers! I was forced to become a disciple of Huber, on the fertility of workers in certain cases, and that they lay drone-eggs only, that is, for the time being, until a new feature was thrown over the subject. About the 20th of June, I had several swarms issue on the same day, and unexpectedly finding myself without hives, I concluded that I might as well take the hive in question, and use it; since it was out of the question, for it to be re-peopled by its present occupants, and I accordingly took it, and used it in a case where *two* swarms had clustered together. I took it just as it was, with its bees, honey and combs, and having put about half of the two swarms into another hive, I immediately put the other half into this hive, and placed the two about a foot apart, so that in case I missed getting a queen in *either* of them, the bees in the hive in which no queen should chance to be, would find the other hive easily and enter it. How great was my surprise to find that a war of extermination was immediately waged against the few bees in the hive containing the drone-brood, and in half an hour, every bee that originally inhabited it, *lay dead upon the blanket,* upon which the hive was placed, *and among the slain was a queen, perfect in size and form!* The question then arose, where did this dead queen come from? If there had been a queen with that portion of the two swarms, that I had forced into this hive, such queen would then not have been killed by them. Had there been more than one queen in this portion of the swarm, then it would have been very natural for one to have been immediately killed by the other; and in such case the bees would have remained contented with the remaining queen; but in a few hours the whole of the bees left this hive in which the queen had lost her life, and joined the other half of the swarms, thus giving conclusive evidence that both queens of the two swarms were in the first hive, and consequently, the small family of bees, that I had considered to be *without* a queen, did actually possess one, and it was her majesty that had perished with her subjects.

In all cases of my experience, I had found that different families of bees, or swarms mix peaceably together, while being hived; hence my surprise at the fight in this instance; but it must have been the existence of a queen among them, and the treasure of honey that engendered so deadly a strife.

When the bees departed from this hive, in which the battle had taken place, not a drop of honey remained. It had all been taken in their honey bags, to deposit wherever a permanent abode should be found.

Huber has stated that queens are never slain by workers in combat, but here is an instance to the contrary, of such a nature, as not to admit of a question, of the queen being killed in the general melee, and by the workers, too.

I recently met with another instance of an attempt on the life of a queen by workers. During a remarkable season of cold, wet and drizzly weather, that lasted about two weeks, some of my bees commenced robbing their weaker neighbors, and one day, while standing in front of one of these invaded hives, watching the destructive strife, I beheld a queen on the ground directly in front of the combatants, struggling with a worker. The worker embraced her, with curved abdomen, endeavoring to find a penetrable point, in which to plant its deadly sting. I seized the queen, but in my anxiety to save her from harm, she escaped and flew away. At evening, I found her in a cluster, near the entrance of the hive, in front of which I first discovered her.

I mention this fact to show that workers pay no respect to royalty, when engaged in a general warfare. In this case, it is probable, that the queen was forced out of the hive, in the conflict that was raging within, and was passively the object of one of the robbers' vengeance when discovered. I say *passively*, for whatever may be the attack upon a queen by a worker, she never retaliates. She never lowers her dignity sufficiently to return a thrust made by a subject, but, as it were, bares her breast and says, "slay me, if you have a heart to do it. I choose death rather than defence." But let queen be pitted against queen, and how changed the scene! The modest non-resisting queen, that tamely suffers death from an unfeeling subject, now rises in her majesty, and with eager and deadly aim, rushes to the combat—the struggle is short, one of the two soon lies in the last pangs of death!

To return to our little family, that met so untimely an end—the dead queen changed the aspect of the case materially, and I was forced to conclude, that instead of the drone-brood being the production of fertile workers, it must have been the work of a queen; and here comes up the question of *retarded impregnation.*

The reader will recollect, that I have stated, that Huber experimented on retarded impregnation, and that he states, that when a queen is retarded beyond the *twenty-first* day of her age, in her impregnation, she lays only *drone*-eggs thereafter, during her whole life!

In the foregoing case, I examined the premises thoroughly, to see what ground I had for taking this latter assumption of Huber, as being applicable to the case before me, and I found much to strengthen me in the belief, and in fact to render it almost certain, that it was an instance of retarded impregnation, beyond a reasonable doubt. In the first place, I found some six or eight royal cells in this hive, that had been constructed the season previous; and since a swarm never constructs any royal cells the first season, unless it be in very rare instances of large early swarms, that *throw off a swarm the same season,* and this swarm not being an early one, and to my certain knowledge, not being in a condition to throw off a swarm at any time during the season, the question arises, *why were these royal cells constructed?*

The probable solution to this query is, that sometime in August or September, the queen belonging to this hive, from some cause, was lost, and the workers availing themselves of their power to replace her, or create another in her stead, constructed the royal cells as above, and reared a queen. This queen coming into existence, at a period when very few drones exist, if any at all, must, from that cause, have found great difficulty in encountering them on the wing, and hence, a *retarded impregnation* is almost certain to result to every queen under such circumstances. In large apiaries, say of fifteen or twenty hives, there is generally some one or two hives that allow a portion of the drones to survive much later than usual, and where such drones do exist, a young queen may, after many flights, succeed in her amours.

DIFFICULTY OF EFFECTING THE IMPREGNATION OF QUEENS AT PARTICULAR SEASONS.

Huber states, that on the occasion of a young queen coming forth at a season of the year, after the usual massacre of drones, he witnessed her ineffectual flights in search of them, for many days; at last she returned bearing evidence of success. This accords with my own experience in similar cases, and I must, therefore, come to the conclusion, that mine was a case of retarded impregnation of the queen, since every fact pertaining to the case, goes strongly to prove it. We account for the great decrease of bees thus:—the fall months of the season were a perfect blank in the increase of this family; hence, when spring came, as a matter of course, we find but a very few bees alive, for the majority of all bees existing in the spring of the year, are brought into being during the fall months previous.

FERTILE WORKERS NEVER EXIST, EXCEPT IN CASES OF A FAILURE TO PRODUCE A QUEEN.

Another circumstance attending the existence of fertile workers is, that *they never do exist, only in cases in which the bees have been unsuccessful in rearing a queen.*

When a queen comes into existence, her natural aversion, and unrelenting animosity towards any thing like *rivalry*, cause her to rush on all other queens yet in embryo, and such workers as have had the misfortune to take a sip of royal jelly, are scented out for immediate slaughter. But when a failure in raising a queen takes place, these poor royal pap workers are allowed to exist so long as no queen is present to immolate them. Thus it will be seen, that the chances of such fertile workers coming under the observation of apiarians, is quite limited.

There is much interesting information concerning the habits and economy of this class of bees, that cannot well be embraced in an especial chapter devoted to workers; but such matter will be unfolded, through the various subjects that I shall consider essential to discuss hereafter, in succeeding chapters. The same may be said of both queens and drones; yet I have thought it best to confine as much matter as possible, in separate chapters, devoted to each respective class—the better to guide the reader in his researches, for any particular information that he may wish to refer to.

CHAPTER III.

DRONES.

THE drones are the largest class of bees in the family. Their bodies are thick, short and clumsy, and obtuse at each extremity. There are two descriptions of males—one not larger than a worker. This class of drones is but seldom seen. How they are produced, is a subject for speculation. It is probable, however, that they only exist, when the queen has deposited a portion of drone-eggs in *worker*-cells; the size of which will not admit of a full development. The common drones are as large as two workers. The head and trunk are covered with dense hairs—much more dense than on workers, or on the queen. Their wings are large, and extend to the full length of the abdomen. Drones have no sting, and may be handled with perfect impunity. They make a loud, buzzing noise when on the wing.

NATURAL USES OF DRONES.

The natural uses of drones have hitherto been a subject on which the greatest contrariety of opinion has existed; especially in Europe. In our own country, those few authors who have written on the bee, have, as I before stated, servilely copied the endorsement of Huber's theory, from foreign works circulated here; that is, in such treatises as have made any attempt to elucidate the natural history of the bee; consequently the question has not been subject to that dispute here, that it has been in England and on the continent.

Huber's theory of the impregnation of the queen, has met with a very strong opposition in Europe, even to ridicule; yet I consider him right—yes, not admitting of a doubt in the mind of any man, who will look into the subject, with a mind untrammelled by prejudice.

The drones *appear* to be a superfluous legion, of no use at all, but rather a disadvantage. This class of the honey-bee, derive their name from their general lazy habits, spending their time in luxury, and feeding upon the stores gathered by the ever industrious workers. They collect no honey at all, for the reason, that nature has not provided them with honey bags, such as the workers possess, to contain collected sweets; neither have they any cavities, or baskets upon their legs, as workers have, to hold pollen or farina. This insect is the only thing known to exist in the animate creation, unprovided with the means of supplying itself with food from the boundless store-house of nature. A drone could not exist a day, were it deprived of the privilege of feeding on the stores of the hive already gathered. They are never seen to alight on any flower, or doing any thing to aid the prosperity of the colony. In one respect they differ entirely from the workers, having the liberty of entering different hives with perfect impunity, while a worker enters any hive but its own, at the peril of its life.

Now, the question is, what are these apparently useless bees for? Would not our apiaries be generally benefitted, could we banish these lazy drones from our hives? This may reasonably seem to be the case, to one not acquainted with the natural history of the bee; but should we banish these bees from our hives, *depopulation would speedily follow.*

CAUSE OF THE EXISTENCE OF SO MANY DRONES.

However mysterious the ways of animate nature may appear, nothing is created in vain. Nature, in order to ensure her legitimate objects of fructification, is ever profuse, often far exceeding the positive requirements of the case, as we may view it; but after all, nature is right and we are wrong. Look, for instance, to the fructifying farina of the tassel of maize, that contains a thousand times the quantity that is necessary to give birth to the ears that brace each stalk around. The captious and precarious winds, that are commissioned to waft this farina to its destiny, are not to be relied upon; hence the vast superabundance that nature has provided to render fertility sure.

Not unlike this is the legion of drones that lazily hang around our hives; and where a *thousand* exist, *nine hundred and ninety-nine* are perfectly useless, save upon the same principle of superabundance, as shown above.

The only object for which drones are brought into existence is the *impregnation of the queen,* and if a less number existed, her fecundity would be jeopardized, in the ratio of the decrease.

IMPREGNATION OPERATIVE FOR LIFE.

Coition is always effected high on the wing, and when once effected, it is operative for an entire season—even during the entire life of the queen. The cavillers at this theory, attempt to cast ridicule on the hypothesis, of a *single* impregnation being sufficient for the natural life of the queen; and, say they, "we admit that if your theory has any ground to stand upon, it is reasonable to suppose, that a single impregnation would suffice for one season, since analogy teaches that; but how do you suppose that an impregnation this spring, effects the queen at her next spring laying; since the winter months are a season of barrenness with her, and certainly no man in his senses would suppose that the coition of the year before, could possibly have any influence on her at that period! You may as well say, that the dung-hill fowl has no need of the male *after the first impregnation,* to render her eggs productive during her whole life!"

All this is reasonable logic, but it avails nothing in the case before us, since there is so much in the history of the bee, that has no analogous bearing with any other similar matter in animate nature, that we cannot rest any theory solely upon such a basis.

We must, and do confess, that if any positive proor can be adduced, showing that impregnation is not effectual, even during the natural life of the queen, then the theory of impregnation with the drones on the wing is untenable. But such proof *cannot* be adduced—on the other hand, it is perfectly reasonable, that the queen should never lose the virtue of a primary coition, because there is seldom, or never, a *total cessation of laying* in the strongest families. I contend that in every strong and healthy family of bees, brood may be found every month in the year, and that the ovary of the queen is never wholly void of the fecundating principle, after once being fully impregnated. I do not say that brood may be found in *every* hive, because half of the hives in existence at the present time, are not in that condition that nature intended a family of bees to be in.

There have been so many tinkers at work, of late years, in forcing bees out of their natural habits, that it would not be surprising if the whole race of bees should become extinct, before the beginning of the next century. Nature so intended a family of bees, that a sufficient body of them should always be together, to be able to generate a natural animal heat even in the dead of winter; and such families, having a healthy queen, will seldom or never be *wholly* void of brood in their tenements. I do not suppose or contend, that in the winter season, the bees are breeding so as to make any material accession to their numbers, even in a state of the greatest prosperity, but a very few larvæ may be found in the coldest weather, in many strong families.

But what are we to do with those families that are weak, and in which the queens discontinue laying in the fall, and do not commence again until the following spring? Such queens have no possible opportunity to have commerce with the drones, and yet they are fertile. Here is undoubtedly a cessation of ovi-positing, for some *four* months. Does the impregnation of the spring previous, operate in this case? It unquestionably does, however strange it may appear. I look upon the question in this light:—that the germ of the ovary, after having been fructified, never wholly loses the efficacy of coition, and though there be a cessation of laying, yet the germinating principle is never lost, but rather lies dormant, until the genial warmth of spring arouses it to action.

If the foregoing premises be fallacious, let us have a proof of its fallacy. They who deny this theory, do not pretend to adduce any theory at all, but rather suffer the case to go by default.

VISIONARY ALLEGED USES OF DRONES.

Some apiarians, however, contend that the drones fructify the eggs as fast as laid, by some means that they cannot well explain, and this is their sole use; but when asked how the eggs that are laid in the spring, *before any drones exist,* become fructified, they acknowledge their inability to answer. Thus is this question beset with difficulties that will probably remain as long as time lasts.

The time when drones appear, as well as the time when they disappear, strongly shows that their use can be no other than the fructification of queens. If their use were for the various purposes that have been ascribed to them, such as fructifying the eggs—feeding the larvæ—sitting on the eggs—producing the necessary heat in the hive, for maturing the brood in due season, &c., how is it, that the brood is regularly perfected, *when not a solitary drone exists?* In the spring and fall, we find the brood going through the different stages, to perfect development; but no drones exist at that time; hence it is time lost to argue this question, with those who advance so unreasonable positions.

I consider the above uses ascribed to drones perfectly chimerical; rather exciting a little surprise, that so palpable errors should be promulgated at this late day, by men professing a scientific knowledge of the nature and economy of the bee.

Another gross error is promulgated, and confidently believed, in many parts of Europe, and especially Poland, in regard to the uses of drones, which is, that they are especially and solely the "*water carriers*" of the family! I have a Polish work on bees before me, making this assertion, as gravely as if the author were promulgating a well known truth, that admits of no refutation, or even question of its accuracy. This, as well as the foregoing uses of drones, are the visionary fallacies of bee-keepers of old times—many centuries ago; and which, with numerous others, as wild and ridiculous as the ignorance and superstition of the times could engender, still exist to a great extent, among the bee-keepers of every country. It were a Herculerian task to eradicate these superstitious traditions—sooner would I attempt to civilize and educate the Hottentot of Africa, than to attempt to *unlearn* the unread bee-keepers of our country, of all their *whims* and traditionary notions, respecting the honey-bee. Their knowledge of this insect is rated by the length of time that bees have been kept in the family; and that man who dates a family possession through several generations, would be a dangerous person to expostulate with on the impropriety of his management, for, it were ten chances to one, that we should receive a *forcible* illustration of the *strength* of his arguments, in the way of ejectment from his premises.

HUISH ENCOUNTERS A SAVAN BEE-KEEPER.

Huish relates an instance of his being introduced to a genus of this species, who had kept bees a long time, and who supposed that he was the veritable "Prince of apiarians;" and on some improvement being suggested by him, followed and backed by argument, he was politely shown the way to the street, in so significant a manner, that it would have been rashness to have delayed the parley.

Nothing will excite the ire of these gentry so much as to question their knowledge of the true science of bee-management. In consequence of this fact, I have ever avoided any controversy with people of this description; and on a recent tour through the State of New York, I made it a point to call on every bee-keeper in my route, that I could visit conveniently, merely to gratify a curiosity that I felt, to see how they generally managed bees. I elicited their management by simple questions, and they generally took great pains to give me all the information in their power; for I never ventured to play the teacher, but humbly and civilly received instruction from them, such as they were able to impart, being a stereotype of the management that was in vogue centuries ago, to a great extent.

I had the pleasure of meeting with many apiarians, who have not despised to read and learn. One gentleman opened the chamber of one of his hives, and to my surprise, drew forth several volumes on the management of bees, which he was accustomed to study, under the balmy shade of the surrounding trees. I found, on the whole, a spirit of inquiry abroad on the subject, and many had been the willing victims in the purchase of a variety of patent hives—not one of which answers the purpose, *as recommended!*

One gentleman said that he would give a large sum of money, if his bees were out of a lot of *patent* hives, and back in his *old-fashioned boxes*; and I found the same desire prevalent among almost every one, who had embarked in patents, to any great extent of time.

WHEN DRONES APPEAR AND DISAPPEAR.

Dr. Bevan says, "the drones make their appearance about the end of April, and are never to be seen after the middle of August, except under very peculiar circumstances."

In my experience, I have found that the drones do not appear, to any great extent, until the latter part of May; and the general massacre takes place in July, and is continued through August. I have this day (August 23d, 1848) seen many drones about my hives, and still under no "peculiar circumstances."

The great slaughter has generally been consummated among the tenants of my apiary; yet scattering drones are found here and there that have escaped an unnatural death. I am fully aware of the "peculiar circumstances," to which Dr. Bevan refers, but I think that author is in error, when he says, that drones are never to be seen after the middle of August, unless under peculiar circumstances. He should have put it *one* month later.

It will be seen that the appearance of the great body of the drones is *coeval with swarming*, and their disappearance as a body, when the swarming season is *terminated.* Now, admitting that their sole use, is the impregnation of the young queens, that issue with the swarms,—is it possible for them to appear at a more appropriate period, or leave at a season that would be better for the prosperity of the colony? Since they must live on the stores gathered by the workers, they should not appear before the time of actual requirement; and when their services can be dispensed with, they should not remain a day to consume the food that is gathered with so much toil and industry. Man, with all his wisdon, could not better this wonderful operation of nature! Had I the direction of the production of drones, I should say, "let them appear in force from the 20th of May to the 1st of June"—precisely the time that they *do* appear. On the first week in July, I should say ' depart"—just the time that the massacre *is* commenced. On the 6th of July, I discovered the first attempt to expel the drones, this season. Thus, nature has ordained this matter; and blind indeed, must he be, who can resist the almost self-evident truth, of the legitimate uses of drones.

"But," say the cavillers, "why should a thousand or more drones be brought into existence, when *one* is sufficient, according to this theory."—It is a true adage, that "none are so blind as those who *won't* see." Thus it is, in the present case. Now, the queen cannot possibly become fertile, without meeting a drone on the wing, in the air. This is her nature, and she may be confined with thousands, yet it is utterly impossible for her, to be fructified by their presence. Then, since she must go forth, and that too, in the regions above, far out of the sight of man, to effect her object; she must not go in vain.

The life of a queen is too valuable to be jeopardized in fruitless sallyings, subject to be caught by the fowls of the air, or to mistake her domicil, on her return, and enter another and perish. A young swarm is solely dependent on the safety of their queen; and if she perish, *ten thousand* subjects die with her.

The great Creator of animate nature foresaw all this, in his infinite wisdom, and wisely created so many drones, that the queen could not well fail, to come within the circle of their flight, soon after leaving her hive, and thus render her fertility sure, on her first exit.

DRONES GO FORTH TO MEET THE QUEEN.

The drones have received a command from the mouth of Him who created them, to "go forth to meet their royal mistress;" and for five thousand years, this mandate has been implicitly complied with. Time may roll on, yet these drones, faithful to the Omnipotent hand that gave them instinct, will continue to take their ærial flights, as regularly as the sun rises and sets.

Perhaps the reader may not have been impressed with the circumstance of the drones, at a certain hour of the day, coming forth from their hives, and taking their flight heavenward! This is a singular truth. Generally from one to three o'clock, P. M., on every fair day, a loud buzzing noise may be heard among the bees. A great commotion ensues, and one is often mistaken, supposing that a swarm is about to issue. This is the general egress of the drones. They ascend in horizontal circles, in an oblique direction; and after being absent an hour or more, return to their hives. This flight takes place daily; and since the drones have no possible cause for leaving the apiary, to gather food, does not their periodical flights, in this manner, show *conclusively,* that nature has bidden them to go forth to meet the queen? Now mark the harmony of the arrangement! The queens, by the same power of instinct, leave their hives about the same time that the drones take their exit; or generally, a short period before, and seldom return unimpregnated.

Huber says, that at *eleven* o'clock, A. M., he witnessed a queen go forth; but I have never yet seen one sally out at that hour.

DANGER OF THE QUEEN BEING LOST, DURING HER EXCURSION.

That all these things should be thus made to harmonize, for the well-being of the bee, is apparent. How easy a thing it would be, for a queen to lose her way, on her return to her hive, if she had to go forth many times.

In her flight, every object that presents itself is new, save what may have been noticed by her on the day of swarming. She sees many hives of the same color and size, and it is only by the most astonishing sagacity, that she is enabled to escape the vicissitudes of a single flight; and were she compelled to go out daily, for any considerable time, not one family in ten would escape destruction; for to be without a queen, is *certain ruin*, when no eggs or larvæ exist in the hive.

HUISH'S VAGARIES RELATIVE TO THE USE OF DRONES.

Huish is a great advocate of the drones being for the purpose of fecundating the *eggs*, instead of the queens. Hear him:—"If by any accident or untoward event, a hive be deficient in drones, the fecundation of the eggs of the queen does not take place, and consequently, no swarms are produced."

It is strange that Huish should make such a declaration as the foregoing, in regard to this subject. Who does not know, that the eggs are fecundated in March and April, long before a drone exists? It makes no difference at all with swarming, whether drones exist or not, as every ordinary bee-keeper knows.

Huish also says,—"When a hive swarms, a number of drones follow the emigrants, in the proportion of the number of working bees."

In regard to this point, it is true, that a portion of the drones in the hive go out with the swarm; the numbers varying, according to the number of drones in it—a mere matter of chance. They go with the swarms from instinct, so as to divide their maintenance more equally among the colony.

THE CONFLICTING OPINIONS AND THEORIES OF OTHER WRITERS DISREGARDED.

I should only be adding mystery to the subject, were I to fill my pages with the conflicting theories and declarations of *Huish, Huber, Bevan, Shirach, De Reaumer, Riems, De Braw, Swammerdam, Hunter, Dunbar, Butler, Thorley, Wildman, Keys, Bonner*, and a score of other foreign writers on the bee; and I think I study the interest of the bee-community, for whom I write, by thus doing.

Dr. Bevan's work is almost entirely made up of the conflicting views and theories of different authors; and when one has perused it, he is about as much in the dark, on many important points, as before reading it.

It is my aim to give a straightforward treatise, without vacillating to the right or to the left, to follow this or that author; but to unfold the truth as it is, and which I have demonstrated, from personal observations. I cannot, however, suffer the fallacies of some of the above authors to go without comment, but I shall be as brief with them as possible.

PARTICULAR INSTANCES OF DRONES BEING ALLOWED TO LIVE THROUGH THE WINTER.

Huish further says,—"Huber says, that he has seen drones in a hive in January, and Mr. Duncan supposes, that they were allowed to remain in the hive, on account of the additional heat which they would generate in winter, or perhaps, they may be preserved for the purpose of pairing a new queen. Those suppositions, however, of Mr. Duncan have not a tittle of truth to stand upon; *not a drone was ever seen in a hive in January.*"

In this case, Huber is right, and Huish is greatly mistaken.

It does not admit of a question, that occasionally a few drones are allowed to winter over, in some hives. What these drones are thus allowed to live for, is a question that will never be answered, so as to cover the whole ground. It is not for the additional heat that Mr. Duncan speaks of, because their numbers are so small, that such a thing is out of the question. A hive is never seen with a *full* complement of drones in the winter. A dozen or so, is the most that I ever heard of, and *four* is the most that I ever found myself.

Last spring, in the month of March, I saw four drones issue from one of my hives.—It is true, I never saw any in January; but those that I saw in March, were in existence in January, and if I had driven out the bees, I should have seen them, of course.

The reason why drones are sometimes left is, that the family is without a queen, or that the condition of the family is such, as to possibly require them to impregnate a new sovereign. If the queen is not in a healthy and sound condition, the drones, or a few of them, are always allowed to exist. This is right—who could order better?

In case of the death of the queen, how important are these drones! In case of the absence of a queen at the time of the general massacre, and no larvæ are left to replace her, the drones are reserved as being needful, in case of the bees being put in possession of a new queen, on which they can make no calculation by any natural means; yet instinct teaches them to preserve the drones, and trust in Providence for a queen.

It may be set down as a fact, that when drones are found long after the general extermination, something is wrong, and needs the attention of the apiarian; but in some cases, it defies the knowledge of man, to conceive why they are left. In the case in which I saw the four drones, mentioned above, the family was in the most perfect prosperity. The queen was very fertile, and I cannot say why the drones were permitted to winter over, unless something was amiss with the queen in the fall, of which she recovered before spring.

For the purpose of impregnating a new queen, a few drones would render the act somewhat precarious, but I presume that instinct teaches them in every emergency, to so act that the end will be effected, for which nature designed them.

THE OLD QUEEN ALWAYS GOES OFF WITH THE FIRST SWARM.

It is not necessary, that the drones should appear in force, until the second swarms issue; for the reason, that the *old* queen goes off with the *first* swarm. This is another point, that has also been disputed; yet it is a fact, that can never be subverted. This is a circumstance, that may well excite our admiration.

Nature is ever careful of the perpetuity of her species of the animate creation; and in the case of the bee, she is not lacking in that wonderful chain of circumstances that produces one harmonious result.

In order to effect this object, viz: the sallying forth of the old queens with the first swarms, nature implanted an implacable enmity between all queens, from the moment of their existence; and even so far, as to force the mother to destroy her own progeny, before it emerges from the cells.

A young queen, that has not been out of her cell, more than five minutes, rushes upon her sisters in royalty, and wrests them from their tenements, while yet in the pupa state, but for the restraint held over her by the workers, who stand in the defence of their young sovereigns, and when a queen approaches, with deadly aim, they seize her, and hold her a prisoner. This natural instinctive hatred of rivalry in queens, is the basis upon which the *rationale* of swarming rests.

In order to arrive at the point, that I had in view, viz: to illustrate the fact of the old queen going off with the first swarm, as briefly as possible, and not run into a chapter on *swarming* at this place, I will simply state, why the old queen does thus leave the hive with a primary swarm.

As soon as the young queens' cells are sealed, or a few days thereafter, say about eight or nine days before the development of the oldest among them, the natural hatred of the rivals, that she has produced, or at least, deposited the eggs in the royal cells for their production, is so great, that sooner than remain to encounter them, she quits the hive, and in her exit, takes a portion of the family with her. If she were to remain in the hive until one or more of the young queens should emerge from the cells, she would, from her superior strength and command over the workers, fall upon such queens, in despite of the efforts of the workers to prevent it, and slay them, without the least compunction or mercy. Nor would she stop there—if any queens yet remained in the embryo state, she would, in her rage, tear off the seals of their cells, and drag them out, as if they were the deadliest enemies to her race. Thus would there be no swarming, since the old queen would not go off, and leave the family without a sovereign, and she will spare none, when once her appetite for slaughter has been whetted. There is much of interest that may be said on the general circumstances of swarming, which I shall endeavor to lay before my readers hereafter.

DRONES SAID TO DIE IMMEDIATELY AFTER COITION.

A remarkable circumstance is said to occur to drones, in their amours with the queen; which is, that death ensues immediately after coition! Since it is impossible that man should ever witness the act of connection between the drones and queens, it is a very difficult question to determine, whether the drone dies immediately or not. However, we have analogy, it is said, in some of the insect tribes, to corroborate this alleged fact.

THE GENERAL MASSACRE OF DRONES.

The general or usual time that the massacre of the drones takes place is, as I have already observed, in the month of July. There may, however, be instances of their being expelled in June, say the last of the month; and there may also be instances of their being allowed to exist until August, before any expulsion takes place. The time of massacre or expulsion is earlier or later, according to the latitude of the location of the apiary. For instance, the expulsion may take place in the latitude of the city of New York, two weeks sooner than in the latitude of Buffalo or Boston.

Strange as it may appear, the manner in which the extermination of drones is effected by the workers, is a matter of contention among the apiarians and naturalists of Europe. Some assert that the bees use their stings, while others contend that the drones are simply disabled, and then cast out of the hives.

Huish says,—"It is the opinion of some naturalists,

that the bee kills the drone by means of its sting, but in the many hundred times that we have witnessed the destruction of drones, we never yet observed that the bee made use of its sting."

Huber is an advocate of their being stung to death; he says:—"On the 4th of July, we saw the workers actually massacre the males in six swarms, at the same hour, and with the same peculiarities. The glass table was covered with bees full of animation, rushing upon the drones as they came from the bottom of the hive; they seized them by the antennæ, the limbs, and the wings, and after having dragged them about, or, so to speak, after quartering them, they killed them by repeated stingings directed between the rings of the belly."

The truth is, that both the way of Huish and also that of Huber, is practised by the bees. In some instances, I have noticed that scarcely any were stung, but the bees cut the cords of their wings, and then expelled them from the hives. Drones that have been treated in this manner, may be seen running to and fro upon the ground, every now and then making a fruitless attempt to rise on the wing. On other occasions, when the patience of the workers has become exhausted, they seize the drones, and curving their abdomen in close contact with their bellies, continue to make their deadly thrusts, between the wings, until successful. In this case, the drones may be seen running around the hive, upon the stand, carrying the workers along with them, which never give up their hold until their object is effected. The workers do seem to have some mercy at times, for long and enduring is their patience on most occasions, in endeavoring to drive the drones away, without doing them any bodily harm. In such cases, the drones quit their usual abode and take refuge in other hives, where, in turn, they meet with the same treatment; finding every hive too hot for them, they return to their original homes, when the workers say, if I may be allowed the term, as the old man did to the boy who was in one of his trees stealing apples.—The old man did not wish to injure the lad, if he could get him out of the tree by the use of moderate means, so he threw a few small tufts of grass at him, and told him that it was wrong to steal apples, and desired him to come down, but this, as the story reads, "*only made the young sauce-box laugh.*"—"Well, well," said the old man, "if neither gentle words nor tufts of grass will do, I'll try what virtue there is in *stones, &c.*" Now the position of these drone-bees is not wholly dissimilar to the above case. The workers wish to get rid of them, indeed, *must* get rid of them. They at first push or drive them off the floor-board by gentle means; finding gentle means ineffectual, they say, "let us try what virtue there is in stings."

The drones, in rushing for shelter from hive to hive, find the best accommodation in those hives in which recent swarms have been placed, and which have not yet been filled with combs. In such hives, they can enter at evening when the bees are clustered above, and congregate on the floor, or bottom-board, huddled together like a flock of sheep, not daring to venture up into the hive. In this manner many nights are passed during the heat of the conflict, and as soon as day dawns, they are again driven out to wander about from hive to hive. In cases of artificial swarms being made, in which the queen is not developed, the bees in such hives give the drones a welcome reception; for the reason that they are absolutely necessary to their prosperity. I had a number of swarms of this kind this season, when the persecution of the drones took place, and on raising the hives in the evening, several hundred drones were found on the bottom-board, as before stated; and it is not unfrequent, that the drones perish from hunger while in this situation. I found *two hundred* drones dead one morning, in one of my hives, in which an artificial swarm had been placed, all lying precisely as they were the previous evening, with their heads towards the centre. In other hives I found many dead drones in the same manner, on different occasions. It is very singular, that every drone in this hive should perish at the same time, but such is the fact; yet not a hair of their bodies had been disturbed by the workers. When I saw this circumstance at first, I was led to believe that some unnatural agency had caused their death; but subsequently finding them dead in the same way in several hives, I attributed the cause to starvation. It was natural to suppose, that this was the cause, since the unremitting warfare made on them generally, gave them no opportunity to partake of any food; for no sooner did one enter a hive, than he was instantly ejected; and the hives in which I found them dead, had not a drop of surplus honey, owing to the unfavorable weather at and before the period of their persecutions. Had a part of their number been dead,—some dying, and others lively, I should not consider it as a singular case; but every bee was dead, and in *precisely* the upright sitting posture, in which they had arranged themselves at evening!

There is, in the circumstances attending the destruction of drones, much to excite our curiosity and surprise–much to reflect on pertaining to the instinctive agency that is brought into action with the workers, at the period when further swarming is known to them not to take place; and the intuition that produces a concert of action, and steels their consciences to all feelings of kindred affection in their merciless ejectment of fellow bees, whose agency has been no less important to the welfare of the community, than that of the executioners themselves,—all of which conspires to elicit the admiration of man, and causes him to exclaim, "verily the wisdom of nature is past finding out!"

One or two points more, and I have done with drones. It has been frequently asserted, that drones have been seen to effect their amours with the queens in the hive, or in tumblers where they had been placed for the purpose of experimenting with them; but the evidence has never been adduced, in so strong a shape, as to be entitled to credence. One thing is certain, which is, that no person ever confined a queen *from birth*, either with or without drones, that *proved fertile*. This has often been tried, but no queen has ever been productive, until she was at liberty to leave the hive, consequently, I consider that the question ought to be forever set at rest, that *the impregnation of the queens is exterior to the hive, and of course while on the wing.*

There are two kinds of drones that may sometimes be seen. I refer to a small black drone that occasionally appears, differing from the ordinary drone, only in color and size. The difference in color may not, in every case, be very perceptible; yet they are generally of a darker hue than the larger drones. This kind of drone is supposed to be bred in those cells that immediately connect the full-sized drone with the worker-cells. On inspection, it will generally be found, that a tier or two of cells exist between the drone and worker-cells, of an intermediate size; and the queen would very naturally be liable to deposit drone eggs in them; and in consequence of the size of the cell, the drone has not room for the natural expansions of his body, and consequently is of less size.

This is undoubtedly the true solution of the question; for it is a well-known fact, that a drone-egg may be put into a worker-cell, by any apiarian, having an observatory hive for experimental purposes, and that it will be nursed by the bees precisely as a worker is treated, and a *small drone* will be the issue. On the other hand, a worker-egg or larva, being placed in a *drone-cell,* comes forth an ordinary worker, not a whit the larger for the capaciousness of its tenement.

It has been said, that drones caress and treat the queen of the hive with great attention and fondness; but those who have made this assertion, are generally the advocates of the impregnation of the queen by the drones *within* the hive, and consequently find it convenient to make out as plausible a story as the case will admit of; but they state, what is not a fact. If the thousand or more drones of a hive, each felt a natural affection for the queen, she would be so harrassed, that she would not be able to attend to her natural duties. The drones pay not the least regard to a queen, any more than to a worker. They remain almost motionless in the centre of the hive, until the middle of the day, when instinct teaches them to depart, as I have already related. This is a wise enactment of nature, in order to preserve harmony within the hive. But no sooner does the drone ascend in his ærial flight, than the instinct of his nature is developed, and he *then* manifests a desire to meet his royal mistress.

It is with reluctance that I feel myself compelled to draw my remarks to a close, on this subject, which I consider one of the deepest interest in the history of the bee, to make room for other matter of importance; and if at any time in the progress of this work, I shall fail to meet the reader's wishes, in the description of any branch of my subject, on the score of general details, I trust I shall be excused, since the subject, in all its various phases and bearings, is too vast for an ordinary volume like this

CHAPTER IV.

EGGS—LARVÆ.—TIME TO DEVELOPE, ETC.

The queen commences laying as soon as tne genial warmth of spring opens. If the weather be very mild, she may commence as early as February, but generally in March and April. She does not, however, commence her "great laying," as it has been termed, until about the first of May. At this period, she deposits from 100 to 200 eggs per day, and as it takes just *twenty* days for a worker to emerge from its cell, fully developed, reckoning the time from the day of laying the egg, it follows that all eggs laid on the 1st of May, will produce perfect bees on the 21st of May.

For a period of about ten years, my bees have not generally swarmed before the first week in June; and the second swarms have issued about the 12th or 15th of June; consequently, those bees that went off with second swarms, must have been produced from eggs deposited about the 20th of May, since a bee is able to leave the hive on the first or second day of its leaving the cell.

DRONE-EGGS—WHEN LAID.

Dr. Bevan says, "the laying of drone-eggs, which is called the great laying, usually commences at the end of April, or the beginning of May."

The great laying of drone-eggs is *always after* the laying of worker-eggs, consequently, I think Dr. Bevan has put the laying of drone-eggs too early, but different climates affect the laying in some measure, and perhaps in England the great laying takes place somewhat earlier than in this country. The bees, of course, have in most cases of swarming, been in existence several days, yet in cases of first swarms, not over a week, and less time in after swarms, and some bees go off the day of their leaving the cells. The appearance of drones takes place in the latter part of May, and in some instances, a few appear by the 15th of May; but I have never found them in large numbers, before about the general swarming season, viz: the first week in June, and since it requires *twenty-four* days for drones to mature from the egg, the great laying of drone-eggs must take place about the 10th of May, in the latitude of New York.

I observed that the laying of drone-eggs always takes place after the laying of worker-eggs, which is a fact; yet when the laying of drone-eggs is over, the queen immediately resumes the laying of worker-eggs, and at the time of her going off with a first swarm, she is ready to proceed with the laying of worker-eggs for some days, when she again commences the laying of drone-eggs but not so extensively as at first.

ROYAL CELLS CONSTRUCTED SIMULTANEOUS WITH DRONE-EGG LAYING.

There is a relation existing between the commencement of laying drone-eggs and the construction of royal cells, worthy of notice.

When the queen has discharged her ovary of its burden of worker-eggs, then she is aware that she will commence the laying of drone-eggs ; for, be it known, that the two kinds of eggs are germinated in perfectly distinct and separate bodies, though no organic separation exists in the formation of the ovary, as has ever yet been discovered. How she knows the fact, that her worker-eggs are exhausted, and that for a few days, she can produce drone-eggs only, is not for me to say ; yet she does know that fact, and the workers know it also ; for no sooner does this crisis arise, than they at once set themselves to constructing drone-cells ; provided, that they be not already constructed, and they build them as fast as the queen requires them, and stop with her termination of laying this kind of egg.

The secret relation between the laying of drone-eggs, and building royal cells is this. The royal cells are always commenced on the occasion of drone laying, when they are commenced at all. It seems to be a signal for the workers to commence this work ; yet, if the hive be large, and only partly filled with combs, not a royal cell will be fabricated. The reason of this is evident, because the bees well know, that they will not have a bee to spare in swarming ; for all their increase will be wanted at home to complete the labors of their own domicile. The bees fully understand their business in all its various branches. No hive ever yet threw off a swarm that was not *full* of bees.

I say, that my bees generally swarm during the first week in June; yet I have had numerous swarms issue in May; and on one occasion, a swarm in the early part of April, which I considered a very remarkable circumstance.

My general remarks on the subject of swarming, with its attendant circumstances, must be reserved for an especial chapter.

THE OPERATION OF LAYING DESCRIBED.

After impregnation, the queen begins to lay in about forty-eight hours. Huber says forty-six, but I have found it to be full forty-eight in most cases, in which I have tested the question. There is no use in being so very particular as to the *hour* and *minute*. No man will care a fig whether it be two hours sooner or later. A description of the operation of laying is correctly given by Mr. Duncan, an English apiarian.

Mr. D. says;—" In the operation of laying, which we have a thousand times witnessed, the queen puts her head into a cell, and remains in that position a second or two, as if to ascertain whether it is in a fit state to receive the deposit. She then withdraws her head, curves her body downwards, inserts her abdomen into the cell, and turns half round on herself; having kept this position for a few seconds, she withdraws her body, having in the meantime laid an egg, The egg itself, which is attached to the bottom of the cell, by a glutinous matter, with which it is imbued, is of a slender, oval shape, slightly curved, rather more pointed in the lower end than in the other."

TIME THAT EGGS REMAIN IN THE CELL.

The egg remains three days before it bursts its integument, and becomes a worm, or *larva;* that is, in natural heat of from 60 to 70 degrees of Fahrenheit, and in colder circumstances the time may be prolonged, even to a perfect suspension of vitality for a long period ; and then, on being subjected to the usual heat, the development takes place in the natural way.

LARVÆ—HOW LONG FED—WHEN SEALED OVER, ETC.

After the hatching of the eggs, which is effected solely by the natural heat of the bees in the hive, generated by the workers, the larvæ are fed from four to six days, according to the heat within the hive, and the cells are then sealed over by the workers, by making numerous rings of wax, commencing at the outside, and finishing at the centre. When the larvæ are sealed over, they commence weaving around themselves a cocoon, or shroud, which requires about thirty-six hours, and from this period until their perfect development, they are called *pupæ, nymphs,* or *chrysalis.* The covering, or seals of drone-cells are quite convex, resembling a half pea in rotundity. The convexity of worker-cells is much less, —almost flat ; and the seals of honey-cells *are concave ;* curving inwardly.

PERIOD OF DEVELOPMENT, ETC.

The period of development of the different classes of bees is as follows, viz :—

Queens from	the	egg,	16	days.
Drones "	"	"	24	"
Workers "	"	"	20	"

The formation of queen-cells, as I have stated, takes place on the occasion of the great laying of drone-eggs in May; the manner of the construction of which is pretty well defined at page 28. The construction of these cells takes place about the 20th of May, and consequently the young queens are ready to go off with swarms in the early part of June.

NUMBER OF BEES IN A HIVE.

Various are the statements in regard to the ordinary number of bees in a hive, and the number of bees that a single queen usually produces in a single season. As regards the number of bees in a hive, it depends much upon whether it be a large or small hive, and whether any swarms have issued from it. Some queens are much more fertile than others, as is the case with the female portion of all animated nature. I suppose the following statement of what an ordinary queen annually produces, to be as near the truth, as we can well get at:

Bees in a first swarm, ...	6,500.
" in a second "	4,500.
" remaining in the parent hive,	8,000.
" produced in the first swarm,	6,000.
	25,000.

In the foregoing calculation, I have made an allowance of 2,000 bees, as being in the parent hive, on the opening of spring; and consequently, 10,000 is the number I compute, as belonging to the parent hive, after the issue of the second swarm. The above aggregate of 25,000 bees from one queen in a single season is moderate. If we take into consideration the number of bees produced by the queen in the *second* swarm, and also that of the queen left in the parent hive, both of which are the indirect production of the parent queen, through her own progeny, we should then swell the grand total to about 40,000 ; allowing the said two queens to produce 15,000!

The above estimate is made on the supposition that two swarms are sent off, and the old queen goes with the first, as she ever does.

If the family had been in a large hive in which no swarming had taken place, the result would have been the same as in the first case; for the reason, that the number of bees sent off in both swarms, viz;—11,000, and the 6,000 that the queen produces with the first swarm, would all have been residents of the original hive, together with the 8,000 produced, and left in the parent hive, according to the foregoing estimate, making in all, as before stated, 25,000. If no swarms are sent off, we lose the 15,000 bees produced by the two queens in charge of them. A queen possesses the power of producing a certain number of eggs in a season; and whether she remain in the parent hive, or sally out with a swarm, it does not affect the aggregate of her laying; provided, that she has room in which to deposit her eggs.

RELATIVE PROPORTION OF DRONES.

The relative proportion of drones and workers is about one to twenty, that is, for a family of workers amounting to 8,000 the ordinary number of drones is about 400. Some writers state the number of drones in a hive to be from 1,000 to 2,000; but they are beyond the mark, as a general rule. There is no law that governs the production of drones, so as to enable the apiarian to make any calculation, in regard to their relative proportion, when compared with the number of workers, that may be relied on in all cases. Some families may have a thousand, while another, equally strong, may have but 500.

Nature does not, in all cases, operate without a loss, or waste of the animal functions; for, in the case of the laying of drone-eggs by the *old* queens, after they have left the parent hive with a swarm, we find *that* brood entirely useless, coming as it does after the swarming season is past.

The old queens are aware of the uselessness of this drone-brood, and consequently, the larvæ are drawn out of the cells and cast on the ground. Why queens are thus compelled by nature, to lay a brood of eggs, that are worse than useless, some one must answer, more deeply versed in the nature of the bee than I am.

It may be said, in the case of the drone-brood produced by queens with swarms, that since a swarm *some-*

times sends off a swarm, that in such a case, drones are necessary; consequently, nature has ordained, that a *thousand* queens shall continue to produce drones, and then cast them out half developed, in order to ensure safety to *one* family that throws off a swarm; for, not more than one swarm in a thousand does cast a swarm the same season. On the whole, this feature of the case appears reasonable; because it is the same principle of nature, that is manifested in the production of 500 or 1000 drones, to render the fertility of a queen *sure*, when a single drone would be sufficient, if that drone could be made, through the instinct of his nature, to be on hand, when occasions should demand his services.

YOUNG QUEENS PRODUCE FEW OR NO DRONE-BROOD.

In the case of a swarm sending off a swarm the same season, it is always a first issue that contains the old queen; and it is she that produces drone-brood; since drones in such a case, would be absolutely necessary to impregnate the virgin queen. But with young queens, the case is very different, and they produce few or no drones during the first season of their existence; but after the first season, they produce the regular number.

POSITION OF EGGS OR LARVÆ.

The position of eggs and larvæ in weak families, where every degree of heat must be economically husbanded, is worthy of remark. In well-peopled hives, the queen deposits her eggs in such locations as may be free from honey and pollen, without any regard to the locality, since the numbers of bees will always admit of generating the proper degree of heat; but when a few workers exist, comparatively speaking, the case is widely different. I have often noticed this circumstance; but a particular instance of this nature has very recently come under my observation. In driving a very small swarm into another hive, for the reason, that there were not bees enough to winter over safely, I found in cutting out the combs, a laying in the middle of the centre comb, about as large as the top of a tea-cup, and about as circular. In the centre were the nymphs or crysalis sealed over; and on the outside of these, were larvæ three or four days old; and exterior to these were larvæ just bursting their shrouds from the egg; and exterior to these were the eggs that had just been deposited. If a needle had been run through the cells of the aforesaid nymphs, larvæ and eggs, it would have passed through cells on the opposite side, containing nymphs, larvæ and eggs of precisely the same age! This is only another evidence of the remarkable instinct of the bee! The nymphs requiring more heat than larvæ three or four days old; and the larvæ of this age, requiring more heat than eggs, how wisely does the bee arrange her broods to the best advantage! In this hive, in which the combs were built, there were not bees enough to allow the least heat to be wasted; and when a cluster of bees is huddled together on one side of a comb, the heat produced is much greater, by having a corresponding number of bees clustered directly opposite. Could human ingenuity devise a better way of economy, in the expenditure of animal heat, in the development of the young of this insect?

CHAPTER V.

DIVISION OF LABOR OF BEES.

HUBER'S theory in regard to the division of labor is, that the workers are divided into *wax-workers*, or those that build the combs, *nursing bees*, and *honey-gatherers;* and he contended, that there is a difference in the organic structure of these different classes, so as to render them incapable of doing anything except the particular labor, that nature designed for them; though such difference in the organic structure is not visible to the naked eye. Huber went a little too far in this assumption, since it would puzzle all his adherents to explain, how such a difference in structure is produced, when they all come from the same kind of egg, and receive precisely the same treatment, throughout their whole development.

But it is true, that labor in a family of bees has its divisions;—there are *wax-workers, nursing bees*, and *gatherers;* but there is not the slightest difference in their organic structure. Man has found, that in extensive laboratories, a division of labor is highly essential: thus, in the manufacture of the *pin*, a single pin passes through many hands before completion.

The builder does not cause his layers of brick to bring them to the place of use, nor to compound the mortar in which they are laid. He finds that each branch of labor, performed by persons for that especial business, best tends to harmony and to a rapid completion of the edifice. The bee, in this respect, is not behind man, in its knowledge of the most effectual application of labor, since it receives its wisdom from a source that knows no error. Man has studied, and found this truth out by *experience*—the bee has this instinct implanted in its censorium from *the day of its birth.*

When man attempts to properly define the beauty and harmony of the domestic labors of the bee, and its wonderful instinctive powers, he is lost in a labyrinth of amazement!

I have, more than once, been inclined to throw down my pen, overwhelmed with the magnitude of the task before me; yet I trudge along slowly, doing but faint justice to the subject, trusting in the charity of my readers, for an exoneration of having failed to meet the case as it merits.

DIVISION OF LABOR PROVED.

When a swarm of bees commence the fabrication of combs in a new hive, a certain number of bees commence the building of them; and another portion go forth to the fields to gather honey and farina; and as soon as the young brood require being fed, a certain number take charge of that duty. This fact, so far as it relates to wax-workers, and honey-gatherers, may be proved in this manner. viz;—remove a hive containing a swarm vigorously at work making combs, to a short distance, beyond the reach of its tenants on returning from the fields, and mark the result. In a few minutes not a single bee will be seen to leave the hive, after such discharging bees have left, that were in it at the time of its removal. Scarcely a bee will be seen to leave the hive during the first day or two after its removal, for the wax-workers are patiently awaiting the return of their comrades that bring in the materials. When it has become evident to the bees that their comrades are lost, (they have no idea of the *removal* of their tenement,) then a new division of labor takes place, and the gathering is resumed with lessened numbers. I have witnessed the above case often, in the formation of artificial swarms from a swarm of such magnitude, that half of its members could be safely spared. The same disorganization of labor is found in the new hive that receives the honey-gatherers only, as they return from the fields; and after a day has past, a portion of the bees that were *gatherers* to the original hive, now become *wax-workers* to the new hive that is placed in the position of the original one, thus proving that all workers are alike, and equally able to lend a hand, at gathering, nursing, or wax-working. The particulars of making artificial swarms, will be given in a future chapter.

POLLEN AND PROPOLIS GATHERERS, ETC.

There is also another division of labor in gathering; for a certain number of bees gather pollen, or farina, which is the same thing, for the food of the larvæ; while others gather honey to store in the cells, and to be used in the fabrication of combs; and, if need be, others gather *propolis*, the wax that is used in stopping up crevices and holes in the hive.

BEES GATHER FROM ONE KIND OF FLOWER ONLY DURING THE SAME EXCURSION.

Again, a division of labor takes place in gathering honey from different kinds of flowers. A bee that commences on the blossoms of the cherry-tree, never leaves that kind of tree for any other, or for any flower, but continues gathering the same kind of honey. So it is with the bee that commences her labors on the apple or pear-tree, &c. In the fields, also, the same flowers are adhered to; and the bee that gathers from the white clover, does not alight on any other flower during that particular excursion! I have witnessed this singular fact, when bees gathering from different flowers came under immediate observation, and almost in contact with each other; yet there was no promiscuous gathering by them.

SENTINELS.

The duty of guarding the hive against the intrusion of enemies, is another feature in the division of labor.

Come when you will to examine a family of bees, you will ever find, at least, one or more sentinels on duty; unless it be in cold weather. If the entrance to the hive be small, but a few bees act as guards; but there they stand, thrusting out their antennæ towards any bee that is suspicious; and let a stranger approach, and there is always some bee on the *qui vive* to arrest its progress. These sentinels are as regularly relieved as those of an army on duty.

THE WONDERFUL OPERATIONS OF VENTILATING BEES!

Last, not least, is the duty of those bees, that in close sultry weather, ventilate the hive, by causing a current of air to be put in motion, by the vibration of their wings. It has often been a matter of surprise with some people how bees can exist in hives densely populated, and having but a very small entrance, that often appears to be entirely closed by the numerous bees around it, when man finds it difficult to find air for free respiration, during the sultry weather of summer; and such persons have supposed that the bee requires little or no air to successfully prosecute its labors within the hive. If such people could witness the indefatigable labors of a large portion of such families of bees, that night and day toil without cessation, to renovate and purify the air within their hives, their minds would soon be changed, and if they were bee-keepers, measures would at once be taken, to admit a little of the pure air of heaven, that is so very essential to their welfare. I cannot better illustrate this subject than to give my observations, in a single case, of the ventilation of a hive by the bees, in my own apiary.

Having a swarm lodged in a hive that I felt particularly anxious should prosecute their labors speedily, in consequence of its being an ornamental domicil, and it being quite late in the season when the swarm was put therein, (22d June,) contrary to my custom, the weather being cold, or rather not warm, for the season, I let the hive down in close contact with the stand, only allowing a few small holes for the egress and ingress of the bees, in order to facilitate the internal heat of the hive. The weather suddenly changing from moderate to extremely hot, the bees clustered in large numbers on the outside of the hive, and their labors seemed almost suspended. On opening the door to the hive, that admitted a full view of all the inside, through a pane of glass, the bees having but partially filled it with combs, I there had a fair and full opportunity to witness the manner in which the bees renew the air of their hives by the vibration of their wings. On the bottom-board of the hive were arrayed files of bees in platoons, as regularly arranged as an army on parade, all with their heads the same way, and keeping up a constant motion of the wings. They were stationed in rows from front to rear, thus giving the laboring bees, that went forth to the fields, an opportunity to pass in and out with the least possible inconvenience; since the avenues between the rows of ventilating bees converged to a focus at the rear of the hive, at which point the bees had built down their combs near to the bottom; and hung there in a cluster around their works, and resting on the bottom-board; and at this point, the bees took their departure, when leaving for the fields, first running along the lanes, or avenues aforesaid, to the point of egress; and those entering, pursuing the same pathway. Being anxious to know what result the letting in of a plenty of pure air, would have on the bees engaged in *ventilating*, I raised the hive on all sides, three-eighths of an inch, and supported it by small blocks at each corner. I then looked into the hive, through the glass door, and saw after a minute or two, the bees commence leaving their stations by degrees, until every column of bees, engaged in renewing the air, disappeared!

CHAPTER VI.

BLACK BEES.

There is a class of bees denominated "black bees," that occasionally appear, and which have caused much speculation among apiarians—some even denying that such a class do ever exist. That such black bees do sometimes appear, is beyond all question; yet many years may pass with the apiarian, without appearing in sufficient numbers to be observed. They are of the same size as the ordinary *workers*, differing in nothing relative to their organic structure, that can be perceived, the only difference being the color, which is a jet black. Huber states that a war of extermination is waged against them, and that they meet a violent death in the same manner that drones are expelled and slain; but this does not coincide with my observations, nor with the observations of any other apiarian, as far as my knowledge extends.

These black bees, when they do appear, which is seldom, are only seen in the summer season, and then in very small numbers. They do not appear to take so active a part, in the labors of the family, as the ordinary workers, and sometimes they seem to do little or nothing. Where they come from, or by what cause they become black, has never been shown. Huber thought that they came forth from the cells black, but it is far more reasonable, to suppose them to become black from age. We know that the very young bee, is of a light grey color; and a few days exposure to the atmosphere turns its color to a darker hue, and old age may cause some bees to become entirely black, at the season of the year when such bees appear. Man's locks turn white with age—some much more than others; and why may we not suppose that age will also change some bees to a jet black, since we positively know, that time does generally give them, in all cases, a darker hue?

CHAPTER VII.

POLLEN, OR BEE-BREAD

EVERY bee-keeper knows what *bee-bread* is; yet every bee-keeper does not know all in regard to this substance, that ought to be known. Bee-bread is the pollen, farina, or dust of flowers, that is gathered by the workers in the baskets, or cavities of their legs—the yellow substance that is carried into the hives so abundantly, in the spring of the year. Bee-bread is the food of the larvæ, or young brood; and the most abundant gathering of it takes place in the spring, when the breeding season is at its height. But this commodity is stored up at all seasons, it being a substance that is not injured by age. In the morning, when the dew is on the flowers, the bees are engaged at this labor, because the dampness of the farina packs better upon the cavities of of their legs, and also that at this period of the day, no honey can be gathered. Here is wisdom!—Man plans his work no better. The bee gathers farina, also when the honey season is past, and when it is not wanted for immediate use. The wants of the following season are cared for, even when the gatherers are extinct, for few live to use the following season, that which is gathered the season preceding.

BEE-BREAD INJURIOUS WHEN STORED IN SURPLUS QUANTITIES.

The gathering of bee-bread at all seasons, though showing forth the indefatigable industry of the bee, is attended sometimes, with serious consequences to the general prosperity of the family. It is in this way:—Bees being ever prone to labor, will sometimes gather a large *surplus* of bee-bread, taking up the room of the hive for years, when there is no possible necessity for its use. The cells that ought to be used for honey and brood, being filled with this substance, lessen the general prosperity of the bees, as a matter of course; and in consequence of this superabundance of farina, the bees require changing from old to new hives, about every four or five years, even if no other cause existed for a removal, which is not the case.

COLOR OF BEE-BREAD—DIFFERENT COLORS KEPT DISTINCT.

Bee-bread is generally yellow, but it may often be seen of a pale reddish hue, and at other times of a slate color. The colors of this substance, as generally gathered, appear to be about the same throughout the world. No change takes place in its hue after being gathered, but it is found of these colors in the nectaries of flowers. A singular circumstance in the packing of this substance in the cells, is worthy of notice. *No two colors are ever found in the same cell!* How the bees are enabled to keep each color separate and distinct, is beyond our pale of knowledge; yet it is but in keeping with their general habits, and regulations in labor.

HOW FED TO LARVÆ.

How this farina or bee-bread is fed to the larvæ is another mystery—that is, whether it be given dry, and in its original state, or whether it be compounded with other substances? No man can ever say of his own knowledge, from ocular demonstration, that a combination of different substances does actually take place; yet collateral evidence does exist, showing plainly that *water* is used in preparing it for use, if nothing more. Water and honey are the only things that apiarians have imagined were compounded with it.

CELLS ONLY PARTLY FILLED WITH POLLEN.

Another singular circumstance attends the packing of bee-bread; it is this:—The cells are never filled beyond about two-thirds of their depth! The remaining space is either left unoccupied, or it is filled with honey. When there is a lack of room to store honey, these bee-bread cells are filled with that substance. Some apiarians have supposed, that the cells are but partially filled with farina, because a covering of honey is necessary to protect it and keep it in good condition. This does not appear to be the case, since a great proportion of the combs containing farina, are generally found to have no such covering of honey.

The cause is probably this:—the bees in traversing the combs, require a foothold convenient, and perhaps quite necessary. In filling the cells two-thirds, or perhaps, three-quarters full, the bees leave a footing; whereas, if every cell were filled to its fullest capacity, of farina and honey—the honey-cells being sealed over, the bees would undoubtedly find difficulty in passing over the combs with the requisite facility.

CHAPTER VIII.

WATER AND ITS USES.

Writers on the management of bees have hitherto given no elucidation of the necessity of bees having water within their convenient reach, beyond the simple assertion, that they either should have water placed daily in pans near the apiary, or that they should be situated near to some stream, lake or river of fresh water. What the effect would be to have no water within the ordinary range of their flight, has never been shown; perhaps for the reason, that an apiary cannot be placed where the bees cannot find fresh water in some place, within the range of their flight, unless it be in a desert.

Even the wells of the neighborhood frequently afford all the water that is required, from the drippings of the bucket, or from the troughs that often stand beside them.

I have often seen bees around my own well, in great numbers, extracting the moisture from the outside of the bucket, or arranged along the gently-sloping sides of a trough, that I had placed there expressly for them. Bees do not like to descend the vertical sides of a bucket, or of any other vessel to obtain water ; because there is danger of falling in ; but a sloping, shallow trough, the sides of which form an angle of from 30 to 45 degrees with the horizon, suits them much better.

HOW FURNISHED TO BEES.

Every bee-keeper should either afford his bees a supply of water at his pump, or well, or place a shallow vessel near the apiary, filled with small stones about the size of a pigeon's egg, in order to give a resting place for the bees, and the vessel then to be filled with fresh water every morning, unless there be a stream of fresh water near, in which case, both modes might be dispensed with. A tin baking pan, about an inch or more deep, is very suitable. Should no stones be put into the pan, many bees would be drowned. I have even known many to be drowned, in cool spring weather, when the stones in the pan were so large, as to admit of spaces or surfaces of water only two inches across! One would suppose that so small a space as this, would be overcome by the bees at once ; and when losing a foothold, and falling into the water, they would paddle across to the stones, and soon take wing again ; but such is not the case in cool weather, such as we generally have from March to June. In *very* warm weather, fewer bees under the same circumstances would perish ; yet water is so benumbing to them, at almost any season, that when once immersed they seldom recover, unless assisted by man, in placing them in some warm, sunny place to dry.

EXPERIMENTAL EVIDENCE OF THE USE OF WATER.

I will now relate what came under my observation, at my own apiary last spring, (1848,) relative to the use and necessity of water in the labors of the bees.

Early in April I placed a tin-pan, filled with small stones, on a bench near my hives. This pan held about a pint and a half of water, when filled with stones. Every morning I filled it with fresh water—sometimes with rain water, and at other times with well water, as it happened to be. I then noted the daily use made of this water by the bees. I had, at that time, but fifteen hives ; yet I found that the pan did not hold enough for them, by once filling every morning. Some days it would be emptied before evening, and on other occasions, the quantity was sufficient for them.

SINGULAR DISCOVERY IN REGARD TO THE USE OF WATER ON VERY WINDY AND WET DAYS.

I *particularly* noticed a very singular circumstance

in regard to the quantity of water taken on very windy days, and also on wet, drizzly days, when the bees could not go to the fields. During such days as the winds were so high, that the bees could not safely go abroad—and we had a few such—the bees crowded around, and into the water pan, in *three-fold* the number they did in ordinary mild, pleasant weather. My apiary had recently been removed to a high and exposed situation, where the winds had a fair sweep ; and on one or two days during the month of April, the winds blew so hard, that the hair on a man's head, almost, I think I may say, required to be held on. I had erected a board fence on the most exposed sides of my hives, to be let down, when the high spring winds had subsided ; and the water-pan being within this enclosure, the bees could approach that without feeling the effects of the blast that swept past them without the yard.

It was on the occasion of the prevalence of such high winds, as before stated, that the bees finding it impossible to go forth to the fields, without being in danger of being dashed to the ground, that they turned their attention to the use of water, to an extent far beyond what they were ordinarily accustomed to use.

THE USE THAT BEES MAKE OF WATER.

Here is a question to be solved, viz :—what use did the bees make of this large quantity of water on those windy days ? One would suppose, that when the general labors of the bees were suspended, that no water at all would be required.

The solution of this question, in my opinion, is this: Bees are wise insects, with a natural instinct that goes far ahead of the brains of man, in many cases. I mean to say, that a large portion of mankind do not possess the genius to adapt means to ends, so well as the little puny honey-bee, so far as its ramifications of domestic economy extend. This being the case, as I presume will be admitted, by all persons acquainted with their general labors, it follows, that the bee studies *economy of labor;* and when the fields cannot be explored, such labor as can be performed, to advance the general prosperity of the family, is undoubtedly attended to.

The agriculturist, when driven from the fields by the storm, says:—" come boys, let us see what is to be done within doors—our potatoes are to be cut, and prepared for planting; or fodder for the cattle and horses should be got ready; the straw cut," &c.

Now, the bee acts on precisely the same principle. Water is used in compounding the bee-bread, and fitting it for the young bees. In the spring, when the weather is cool, a few days' consumption can safely be made in advance, and it is thus that I account for the more abundant use of water on such occasions, as do not admit of the usual labors of the family being performed.

WATER USED IN WET WEATHER ABUNDANTLY.

Not only in *windy* weather, but also in *rainy* weather, do bees make use of a more abundant supply of water than usual. I have noticed almost the same rush to the water-pan, on a damp day, when it did not rain enough to keep the bees confined to their hives, that took place on a windy day; even when every plant and leaf was studded with rain-drops. I was somewhat surprised, that the bees should take the water from the pan, when it could be obtained in a thousand other places with the same facility. The same reason that caused the greater use of water in windy weather, led to the use of it more abundantly in wet weather; and the reason why the bees preferred the pan to other places, in obtaining it in wet weather, I presume is, that the liability to get wet is more when alighting in promiscuous places, when *everything* is wet, than when alighting on the stones in the pan.

DECREASE AND FINAL TERMINATION OF THE USE OF WATER.

The use of the water from the pan continued through the months of April, May, and a part of June, when a great decrease took place in the use of it; and this decrease in the use of water was coeval with the decrease *in the production of larvæ.* Finally, in July the bees frequented the water-pan so little, that I considered it useless to fill it daily, and omitted to pay any further attention to it.

That the bees use water in preparing the food, (farina, or bee-bread,) for their young, is apparent, from the fact, that when breeding declines, the use of water diminishes.

Now, from the foregoing remarks, it appears that water is a much more important article in the economy of the bee, than it has hitherto been considered to be; and how far the bees are benefitted in their general prosperity, when they have an easy access to it in the spring, is impossible to truly define; yet there is no doubt that they are greatly benefitted thereby. The case that came under my observation, as above stated, shows that the time was not lost when too windy or too wet to go forth to the fields; but it would have been lost, had there been no water placed, especially for the use of the bees, in close proximity to the apiary.

A CLOSE FENCE AROUND THE APIARY NECESSARY IN CERTAIN CASES.

It is clearly shown in the foregoing remarks, that where the apiary is placed in a high situation, where the winds meet with nothing to break their force, a board fence around it is indispensable—not too near, but sufficiently so to break the force of the winds. Had I not had such a protection, the bees could not have come out for water, on the aforesaid windy days; therefore, let every bee-keeper, having a large apiary, afford his bees a pan of water in April, May, and June; and those having fewer hives should do the same; unless the bees can get water in the immeaiate vicinity.

CHAPTER IX.

SALT—HOW TO BE USED.

VARIOUS are the benefits ascribed to the use of salt, by the bee-keepers of our country, who profess to have no further knowledge of bees, than that which has been taught them from tradition, or from such experience as they have had in the management of bees, which amounts to letting them take care of themselves, and if they live—well—if they die—it is the same. This is about all the knowledge, that the majority of the bee-keepers of the world over possess.

SALT PUT UNDER THE EDGES OF HIVES.

Salt, say these sapient bee-keepers, should be placed under the *edges*, and perhaps under *the whole hive*, as I have seen many instances, to *prevent the moths entering!* This is a perfect fallacy. No quantity of salt ever yet kept a moth *out* of the hive. The moth is a *winged* insect, and enters the hive, without coming in contact with this salt, even if there were a peck of it there. The moth alights on the outside of the hive,—runs in through the entrance, on the *upper* side generally, and turns directly upward, without touching the bottom-board at all. When the worms are produced from this winged moth, they creep down the side of the hive, and search for a hole or crevice in which to wind up in a cocoon, from which a winged moth issues in a few days, to take its turn at entering the hive, if it can. The salt placed under the corners, or edges of hives, as tradition recommends, from time immemorial, will keep the worms from winding up in a cocoon, under the edges of the hive, where this salt is placed, but the worms have only to crawl entirely out of the hive, and in most cases, they will find a convenient crack or nook to suit their purpose, close at hand. Hence, it follows, that so long as these worms can find *any* place about the hive, to wind up in, the salt placed under the hives is of little or no use, since a moth leaving its cocoon a rod from the hive, is just as able to gain admission, as one emerging from a cocoon directly under the hive; for, if the bees are not strong enough in numbers, to protect themselves in the one, they are not in the other case. Even if a place cannot be found to wind up in above ground, these worms will go below the surface of the earth for this purpose; but it is a last resort, or forlorn hope for them, in such cases, and few winged insects are produced by them in such instances. The true policy of bee-keepers is, to keep everything so snug and close around their hives, as to preclude the possibility of the worms finding any winding-up place; *then* place salt under the hives, and a good result will follow. It is a very difficult thing to place hives in such a position, that no winding-up place shall be afforded to moth-worms; yet it can be effected, and I will hereafter show how it can be done. This chapter is on the use of *salt*, and I cannot inform you at this place, but I will do so, when I come to a subject to which the construction of bee-stands, &c., legitimately belongs.

SALT NECESSARY FOR BEES.

The question, "is salt necessary for bees?" is asked a thousand times annually, in every State in the Union. That is, is it of any benefit to place a lump of salt within their reach?

I answer *it is*. My reasons are simply these: Every thing in animate nature, that seems to desire the taste of salt, it is beneficial to. The cow and the sheep can hardly do without it, as well as many other animals; and it seems to be necessary, in a greater or less degree, to all animated nature. The dung-hill fowl craves it to such an extent, as to jeopardize its life by partaking of too much, when an opportunity occurs. I once lost about twenty young fowls, in consequence of emptying into the barn-yard a pork barrel which contained a few quarts of salt; and I also lost a favorite pet canary bird, by allowing it to come out of its cage, and peck the salt standing on the dinner-table. But bees will not hurt themselves by the use of salt. A lump placed near the hives, under cover, will do no harm, and since the bees will occasionally partake of it, we should judge, that it is best to give it to them.

I do not consider it of much consequence, whether

salt is given to bees or not. I have stated my own views on the subject, and leave the matter to the option of the reader, or bee-keeper, to use it, if he pleases.

CHAPTER X.

PROPOLIS.

HERE again, we broach a disputed subject, viz ;— whether propolis is a *natural* or a *manufactured* substance? It makes no difference to us, in the prosperity of our bees, to know whether it is the one or the other; yet we all have an inkling of *Yankee inquisitiveness* to know more than is absolutely necessary of every subject in which we feel an interest; and methinks I hear some curious "Jonathan" exclaim—"I wish I knew what it's made of!" Well, perhaps you can *guess*. If you can, you can do what has never yet been done.

But it may be well to inform the reader, what substance is meant by the term *propolis*. Propolis is the glutinous substance, that is used by bees to fill all cracks and crevices about the hive. It is much darker than wax, the substance that the combs are constructed of, and it is of a more adhesive, tenacious nature. The quantity that a family of bees sometimes produce is astonishing.

I have some singular and interesting remarks to make, on the use of this substance in particular cases, that came under my own observation, which will come in under the head of the "instinct" or "sagacity" of bees.

HUBER'S OPINION ON PROPOLIS.

Huber considered propolis to be a positive genuine production of nature, and not manufactured, but collected by the bees from the leaves and branches of certain shrubs and trees, the principal one of which, some apiarians consider to be the *tacamabac.*

Huber's opinion on this subject does not, by any means, set the question at rest. Neither he nor any other person, it is probable, ever saw the bees in the act of gathering this substance; nor even when gathered by them, on their return to the hives to deposit their burdens. The first appearance of this substance is at the places where it is used, and since we never see the substance gathered, and know of no shrub, plant or tree, that exudes any precisely such adhesive material, we have no *positive* proof that it is a natural substance; still, it seems that it ought to be gathered from shrubs or trees, since so many trees do send out an exudation of analogous features; but if obtained from trees, why do we not witness the bees returning loaded with it, as they do with pollen? The fact that we do *not* see the bees thus returning loaded with it, almost sets the question at rest, on the score of its being a *natural* substance.

According to Huber, the bees have been observed to

draw out long threads of this viscous substance from the exudations of trees, and to lodge them in the cavities of their legs, and as soon as one bee had completed its load, another bee was very conveniently at hand to continue the same process, until a sufficient quantity had been obtained. They then began to knead and work it like an Irish laborer compounding a heap of mortar, and when in a proper state of attenuation, they proceeded to line and solder their cells. By the way—*lining* and *soldering* cells with propolis, is not a branch of labor belonging to the architecture of the bee at all. Cells are made perfect with wax alone, and Huber has here made an assertion that proves him not to be entitled to what has been awarded to him, viz: the reputation of being an accurate apiarian.

PROPOLIS AN ELABORATED SUBSTANCE.

It is probable, however, that propolis is an elaborated substance—to say that it is *positively* so, would be presumption; for it is barely a possible thing, to discover the bees in the act of producing it, even admitting that they do make it; so impenetrable to the eyes of man is much of their labors, when clustered in darkness.

Some apiarians consider it an elaborated substance, of the same nature as wax—even wax itself, but a little more colored; yet this hypothesis cannot be correct, since wax and propolis are so different in tenacity and color, that they cannot be one and the same thing. If it be elaborated, and of a separate and distinct nature, then arise difficulties as to the *modus ope-*

randi by which it is produced; and here the question must forever rest. The bees *produce* it when it is required, but *where* they obtain it, or *how* they make it, must be a secret not for man to unfold, Huber's assertion to the contrary, notwithstanding.

It has been asserted by apiarians of considerable dis tinction, that propolis is used in laying the foundation of combs; but this assertion is at variance with my experience. The *first* rudiments of new-made combs have come under my observation so many times—even the very first *beginning* of constructing combs, up to every stage of their prosecution, that what my own eyes have so often beheld, cannot be controverted by the statements of all the apiarians of Europe, should they declare that propolis is used to lay the foundation of cells. I have ever found the first rudiments of cells, to be composed of *wax*, the same substance that the *entire* combs and cells are composed of—not the slightest difference in construction and color, could I ever discover.

CHAPTER XI.

WAX.

Wax is the substance of which the bees construct their combs. This is not an elementary or natural substance; but it is produced by elaboration. The most universally-acknowledged theory of the production of wax is, that it is an exudation from the abdomen of the bee, through the openings of the scaly rings which compose that portion of the honey-bee, and that honey is the only original substance from which it emanates. This is truly a wonderful theory, and without a perfect knowledge of the economy of the bee, relative to comb-making in particular, one might be justified in casting ridicule upon it; but when we take into consideration all the circumstances attending their labors, in this department of their duties, we find abundant evidence that honey is the original substance from which wax is produced, and that its elaboration takes place within the bee, coming forth in strings of pearly whiteness.

HONEY AND POLLEN THE ONLY SUBSTANCE THAT BEES GATHER.

The only material that bees are known to gather is

honey and *pollen*. No other substance was ever seen to be brought in by them; and the consequence is, that *wax* is either made of one or the other of these two substances. There is no mistake on this point. Now, let us consider what ground we have for supposing that wax is formed from pollen or bee-bread. Firstly: pollen is only known to be placed in the *cavities* of bees legs—not taken into the stomach of the bee, in its original state at all. Secondly: pollen is known to be the food of the larvæ, and the manner of gathering this article shows conclusively to most apiarians, that this is its *sole* use.

APIARIANS CONTEND THAT WAX IS MADE OF POLLEN.

There are a few apiarians who contend that wax is elaborated from pollen; yet after a careful perusal of their arguments, I do not consider that the evidence adduced by them is of such a character as to substantiate their theory, by any means.

Pollen is gathered in the months of April, May and June, in the largest quantities. At this season breeding is at its height, and consequently more of this substance is required. In some hives, during these months, comb-making is carried on extensively; for instance, when the bees commence labors in the supers or chambers of their hives, or in cases in which the whole interior of their permanent domicil is not yet filled with combs; but in no instance did I ever know of a family of bees gathering a particle more pollen on account of such comb-building, or working in wax being carried on, and I have given the subject my most faithful attention.

I now affirm, that a hive well filled with combs and bees, having no *extra* room for wax-working, may be placed along side of a hive, having the *same number of bees*, but the hive only *half* filled with combs, and the pollen gathered by the bees of the hive that is filled with combs, shall even *exceed* the quantity that is gathered by the bees of the other hive, which shall be vigorously working in wax in filling their domicil with the usual combs. Now, if pollen were the constituent principle of wax, the case would be reversed, and more pollen would be gathered in the hive but partially filled with combs, than in the full one.

POLLEN ADMITTED TO BE A COMPONENT PART OF ORDINARY BEES-WAX.

Again, pollen being a dryer substance than wax, and containing but few adhesive properties as it is brought into the hive; and the color of wax always being white, while pollen is of various hues, seems to put the question at rest, proving that wax must be made from some other substance. It is true, however, that pollen forms a part of wax; when the combs are immersed in boiling water for the purpose of extracting it, the pollen then gives it its yellow hue; but ordinary *bees-wax* is quite a different substance from that which is used to build combs, and the difference arises from the fact that pollen composes a large proportion of this latter substance, when prepared for market; whereas, the original substance used in comb building, is wax in its purity; and this original wax is much superior to ordinary bees-wax, for the purposes for which this latter substance is used. Combs, when first built, will melt down to pure wax, without any waste from impure substances, and it is much whiter and better in every respect, than ordinary wax sold in the market.

BEES, WHEN SWARMING, GO LADEN WITH HONEY.

When a swarm of bees issue from a hive, it is a well-known fact, that they carry with them as much honey as their honey vesicles or bags will contain. I have often known the boxes in the chambers of my hives to be emptied of their contents, during the night previous to the issuing of swarms from the same hives. When this circumstance is noticed in the morning, viz: the emptying of the cells of the supers or boxes in the chambers suddenly, it is a sure sign that a swarm will go off on that day, if the weather continues favorable; yet it is not an easy matter to make such a discovery, since the bees remain closely packed in the chambers, up to the very moment of sallying forth.

The object of the bees in thus going forth laden with honey is, to have wherewith to sustain life for several days, and thus be prepared to withstand any unfavorable change of weather that might intervene before a supply of provisions could be secured. Like the traveller who starts on a journey across some desert waste, not only taking provisions for the journey, but also taking a supply to provide against any reasonable contingency that may retard his progress to the land of plenty whither he is bound.

A FEW BEES JOIN THE SWARM WITH PELLETS OF FARINA.

When the swarms thus issuing from their tenements are hived, perhaps a dozen or more bees may be among each emigrating family that carry with them from the old hives, pellets of farina. That there are any bees among them with such pellets, is a matter of chance merely; for it often does happen, that not a solitary bee thus laden goes off with the swarm.

NO POLLEN GATHERED THE FIRST DAY OR TWO AFTER SWARMING.

Now, the bees, in most cases, commence comb-building within an hour after being settled in their new home, and during the first day, at least, no *pollen* is brought in; still, if the bees be dislodged after twenty-four hours, large sheets of new combs will be found constructed. The question then is, what do the bees make these new combs of? It cannot be *pollen*, for the quantity of that substance carried along with the issue of the swarm, would not construct a half-dozen cells at most, and more likely not a single cell. Nothing except honey is brought into the hive during the first day or two; still, the comb-making goes on the most rapidly from the beginning. It follows, of course, that *honey* is the elementary principle of wax.

CHEMICAL CHANGE OF HONEY TO WAX.

The chemical change that honey undergoes in the

stomach of the bee produces it in its proper state for working. I say *chemical* change, for the reason that the honey being probably combined with some other fluid natural to the body of the bee, and both substances exposed to a gentle heat, produce virtually and truly a chemical change. Nor is it in the power of the bee to stop this chemical change, if the honey remain in the vesicle over a certain length of time, say over four or six hours. No person, to my knowledge, has ever before ventured to make this declaration; yet, if we look properly into this subject, we are forced to this conclusion. How often have bees commenced the construction of combs upon the branch of the tree where they clustered on swarming, on occasions when they have been neglected by their owner, or have not been discovered by him! The bees certainly must know that they could not exist in such situations long, and it would be contrary to their well-known habits in the economy of labor, and their wonderful instinctive wisdom, to build combs where they could be of no use, if they could avoid so doing.

I do not say that the bees cannot possibly avoid building combs under such circumstances; yet I say this; that if the honey-bee fills its honey-bag with honey, and finds no place in which to store it within a certain time, under *twelve* hours at most, the honey thus placed in the vesicle of the bee, undergoes a *chemical change*, over which the bee has no control;—that the new chemical substance, which is wax, exudes through the *scales of the abdomen*, that lap over each other like the scales of a fish, and is taken therefrom by the bee, in threads or strings, and at once made into combs, or it is cast away. That such exudation is ever cast away by the bees, we have no evidence; yet they can cast it away if they please, consequently, I cannot say that the bees cannot *avoid* making combs when their honey-vesicles have been filled. There are instances in which bees are known to remain twenty-four hours in a new hive, without working at all in wax; but in such cases, it is probable that they were not provided with any more honey than just enough to sustain life.

EXPERIMENT SHOWING FURTHER PROOF THAT WAX IS PRODUCED FROM HONEY.

In order to show more conclusively that wax-working is carried on, without the use of pollen, or of any substance except honey. I will narrate an experiment that took place last October.

I had a couple of weak swarms that had gathered no honey beyond their daily supply, and had built but a few short combs. Their numbers were so small that I had no hopes of their being able to survive through the winter. On going to the apiary on a pleasant day, about the 20th of October, I was surprised to see a swarm of bees in the air. They soon clustered and formed a bunch about the size of a quart measure. I found this to be one of the weak swarms before mentioned, that had left its original tenement for some uncertain destiny.

I took a new clean hive, and having, with the aid of

melted bees-wax, fastened a few pieces of clean, new combs in the hive, and saturated them with honey, I then hived the bees, and set the hive in a new location, and fed them plentifully with pure honey. The next day, another swarm deserted, of the same character; leaving a little brood and no honey. I hived them also, precisely in the same way, and fed both swarms with as much pure honey as they could consume, or carry away. I found that both swarms began to build combs rapidly, it being very warm weather for the season; but not a solitary pellet of farina was brought into the hives, as I could discover; and none being in the combs that I fastened in myself, how can it be possible that wax is formed from any other substance than honey? I think my own experiments have settled the question, in connection with the general economy of the bee in wax-working, that has come under my own observation—that is, so far as my own opinion on the subject is concerned, but lest some of my readers should still require further proof, I will now give the experiments of the "Prince of apiarians" on this subject, as a quietus.

THE EXPERIMENTS OF HUBER, SHOWING THAT BEES WORK IN WAX WHEN CONFINED, AND FED ON HONEY OR SUGAR ONLY.

He says: "The existence of the organs before described, and the scales seen under different gradations, induce us to believe them appropriated for the secretion of wax. But in common with other animal and vegetable secretions, the means by which this is accomplished appears to be carefully veiled in nature.

Our researches, by simple observation, thus being obstructed, we felt it essential to adopt other methods for ascertaining whether wax actually is a secretion, or collection of a particular substance.

Providing it were the former, we had first to verify the opinion of Reaumer, who conjectured that it came from an elaboration of pollen in the stomach, though we did not coincide with him in the opinion that bees then disgorged it by the mouth. Neither were we disposed to adopt his sentiments regarding its origin; for, like Hunter, it had struck us that swarms, newly settled in empty hives, do not bring home pollen, notwithstanding they construct combs, while the bees of old hives, having no combs to build, gather it abundantly.

We had, therefore, to learn whether bees, deprived of pollen for a series of time would make wax, and all that is required is confinement.

On the 24th of May, we lodged a swarm which had just left the parent stock, in a straw hive, with as much honey and water as necessary for the consumption of the bees, and closed the entrance so as to prevent all possibility of escape, leaving access for renewal of the air.

At first, the bees were greatly agitated; but we succeeded in calming them by carrying the hive to a coal-dark place, where their captivity lasted five days. They were then allowed to take flight in an apartment, the windows of which were carefully shut, and where the hive could be examined conveniently. The bees had consumed their whole provision of honey; but their dwelling, which did not contain an atom of wax when we established them in it, had now acquired five combs of the most beautiful wax suspended from its arch, of a pure white, and very brittle.

We did not expect so speedy a solution of the problem; but before concluding that the bees had derived the faculty of producing wax from honey on which they fed, a second experiment, susceptible of no other explanation, was necessary.

The workers, though in captivity, had been able to collect farina; while they were at liberty, they might have obtained provisions on the eve, or on the day itself of their imprisonment, and enough might have been in the stomach or on the limbs to enable them to extract the wax from it that we found in the hive. But if it actually came from the farina previously collected, this source was not inexhaustible; and the bees being unable to obtain more, would cease to construct combs, and would fall into absolute inaction.

Before proceeding to the second experiment, which was to consist in prolonging their captivity, we took care to remove all the combs they had formed in that preceeding. Buernens made them return to their hive, and confined them again with a new portion of honey.

The experiment was not tedious. From the evening of the subsequent day we observed them working in wax anew; and on examining the hive on the third day, we actually found five combs, as regular as those they had made during their first imprisonment.

The combs were removed five times successively, but always under the precaution of the escape of the bees from the apartment being prevented; and during this long interval, the same insects were preserved and fed with honey exclusively. Undoubtedly, the experiment, had we deemed it necessary, might have been prolonged with equal success. On each occasion that we supplied them with honey, they produced new combs, which puts it beyond doubt that this substance effected the secretion of wax in their bodies, without the aid of pollen. As the reverse of the preceding experiment would prove whether the pollen itself had the same property, instead of supplying our bees with honey, we fed them on nothing except fruit and farina. They were kept eight days in captivity, under a glass bell with a comb, having only farina in the cells; yet they neither made wax, nor were scales seen under the rings. Could any doubt exist as to the real origin of wax? We entertained none."

Huber also tried the result of feeding on sugar, instead of honey, while the bees were confined. The bees produced wax sooner, and in greater abundance, than when fed on honey.

A pound of refined sugar, reduced to a syrup, and clarified with eggs, produced 10 drams, 52 grains of wax, darker than that extracted by the bees from honey. An equal weight of dark brown sugar produced 22 drams of very white wax—the like came from maple sugar; that is, two ounces and three-quarters was the greatest quantity of wax obtained from a pound of sugar.

Having now given the reader a brief view of the preliminary features of my subject, I think he is enabled to advance to the more interesting part of the work, and to fully understand the merits of the case. I say "to advance to the more interesting part, &c.—I mean to the *practical* apiarian, whose sole object is not amusement.

PART SECOND.

CHAPTER XII.

REMARKS.

A CONSIDERABLE portion of this work will now be devoted to the practical management of bees; since it is this branch of my subject, that will most interest the bee-keepers of this country, at least.

I cannot say, that the rules that I shall adduce will divest the management of bees of all the difficulties pertaining to their successful culture, but such as I shall offer, will have the merit of being the result of my own experience, and not the result of other men's brains.

That a thorough practical treatise on this branch of rural economy is much needed by the American apiarian, I believe every one who is fully acquainted with the subject will admit; but do not understand me, as arrogating to myself the honor of having in this treatise furnished exactly the work that is so loudly called for. I have made an attempt at furnishing something better than the American works extant on this subject—how much better, or how much worse, than my predecessors, I leave the public to judge.

The American treatises on the honey-bee, up to the present period, are quite brief, and very meagre in practical details, as I have before observed, hardly giving the subject even a tolerable discussion in its various branches, and leaving out entirely very many things of the greatest practical importance. I do not thus speak of these works in a spirit of disparagement, in order to build up my own reputation on the wreck of other writers; but I speak as a faithful historian, seeking no advantage save what *truth* will honorably award me.

The various foreign treatises on this subject that have had a limited circulation in this country, are not, in my estimation, suited to the wants of American bee-keepers. Many of them are works of great merit, yet I think the most of them neglect the *practical*, and devote too much space to the *theoretical* portions of their works. Again, they are too expensive for us. Dr. Bevan's work, however, is an exception. His treatise was reprinted some years ago in this country, and sold at a very moderate price, but I believe it is now entirely out of the market.

The art of managing bees in this country is probably as little understood as any other branch of rural economy; that is, so far as profit, health and productiveness are concerned.

It is generally supposed, that bees require little or no care, and if they prove unproductive, or are destroyed from the ravages of the bee-moth, it is a mere matter of chance, wholly beyond the control of the owner.

This is a gross error. The same care and expense that a farmer bestows on his pigs or his poultry, would produce much larger profits if bestowed on the culture of his bees. But bees are not to be looked after or cared for. When their owner passes the hives, he barely condescends to look at them, as if they were crying out, "*noli me tangere*!"—Stand off!—Keep your distance, sir!" This is not right. Every bee-keeper should cultivate a better familiarity with his bees, and know at all times their condition and their wants. The time necessary for doing this is comparatively trifling.

Indeed, the cultivation of bees may not only be made a source of moderate profit in all cases, but when properly attended to, a fortune might be accumulated from the labors of this insect alone!

CHAPTER XIII.

HIVES.

DURING the last twenty years, many new and useless bee-hives have been palmed off on the ignorant and too confiding bee-keepers of our country. Men who have had sufficient brains to devise some new plan and style of hive that did not before exist—who never understood a single principle of the management of bees correctly, have, by dint of unblushing falsehood and impudence, bled the bee-keeping community pretty freely. Let it be here understood, that whatever I may say in the following pages in regard to different hives in use in this country or in Europe, I shall be actuated by no other motive than simply to relate the truth, as I view the case. Every person acquainted with the subject, will bear me out in the assertion, that there has never yet been introduced or invented, a hive that, after a patient and fair trial, ever answered the purpose as originally recommended. The reason is, that inventors promise *too much* —more than it is possible for the bee to accomplish. It is true, that I shall introduce a hive of my own in this work, and many persons may suppose, that the pungency of my remarks upon other hives, is to open the way to my own; but in this they are greatly mistaken, for the whole tenor of this book is against *all* hives promising great and unprecedented results in the gathering of honey, &c., and had it not been for the fact, that some one is now selling a hive as a *patent of his own invention*, got up by me some years ago, and gratuitously tendered to the public through the Am. Agriculturist, I should not have brought forward the hive referred to, except as free. Had I wished to make the greatest possible amount of money out of the sale of my hive, without any conscientious scruples reining me within the pale of truth in my declarations of its merits, I should have written a very different book from this; and the reader will at once see, on perusing this treatise, that my general remarks on hives are anything but such as will aid the sale of my own.

SIZE OF HIVES.

The first desideratum with the apiarian is, the proper dimensions of hives. As the builder in rearing his edifice, sees that its foundation is firmly laid, that the superstructure may not be impaired; so does the apiarian look to the correct size of his bee-hives, that his subsequent labors may not prove in vain, in the management and culture of his bees.

Notwithstanding the enquiry has been abroad throughout all christendom during centuries, in regard to the true shape and size of bee-hives; yet we stand in the same position that we did a hundred years ago, relative to this important question. Every bee-keeper has his size and shape, and no one is able to set the question at rest. We find hives from the little box of six inches square, made expressly for very small swarms, up to almost any dimensions, even to the size of a barrel. There seems to be a perfect chaos existing in the minds of men on this subject, or rather, that every man's views on this subject, are so vague and undefined, that a chaotic confusion is the general state of public sentiment on this very important branch of bee-culture.

Now, can any one reasonably suppose that there is no solution to this query? Does any one presume, that a small hive or a large hive; a short hive or a long hive, is all the same; making no difference at all in the general prosperity of the apiary? No one can thus think, for it is contrary to the general principles of common sense; yet bee-keepers, to a great extent, *act* on this principle, and fill their bee-gardens with every manner of hive, throwing all *system* to the winds.

As I look upon this subject, there must be a *right* size, and a *wrong* size—a *right* shape, and a *wrong* shape. But the grand question is, what is the right size and shape? There's the rub! Who can answer? In my opinion, every bee-hive in the United States should be of a certain size and shape.

SPACE NECESSARY FOR SWARMS.

The queen is able to prodúce a certain number of larvæ, or brood, in a season. She requires a certain area of space in which to deposit her eggs, and more than enough is worse than useless. Like the coat upon one's back, a close fit is required; beyond, or short of this, is either ruinous, or highly disadvantageous.

It is true, that some queens are more fertile than others—even the same queens produce more larvæ some seasons than in others. This is quite natural, since a bee is liable to be affected by various vicissitudes of life, as well as any other animate being. But admitting this, we then wish to know what space is necessary to afford an average area for a queen's use, giving, as a general rule, as much room as can be used, and at the same time leave no waste space?

NUMBER OF WORKERS ADVANTAGEOUSLY EMPLOYED.

Again, how many workers can be employed in the same hive to advantage? There is an answer to this—a definite answer; yet I never beheld the subject mooted, as well as many other important questions, touching the management of bees, in any work published in either Europe or America.

The case lies simply here :—you may put a queen into a hive suited to her requirements, and you may then give her just as many workers as she ought to have ; that is, the number that will readily construct the required complement of combs, and have the various branches of labor pertaining to the family all progressing harmoniously, without any branch being retarded, to the detriment of other branches. You may then add to this specific number of bees that constitute just enough, a few more thousand, and you derange all their labors, by an *excess* of laborers. Every one knows, that when a body of mechanics are at work on any kind of employment, and as many are thus employed, as can conveniently find room to labor in, that if another body of men be thrust in to aid them, that instead of being an aid to them, they would actually retard the work. So it is with a family of bees. When once a family have *enough* laborers, more are worse than useless, and they retard the labors of the family, by crowding among the combs, and also farther injury is done, in consuming the stores in a greater ratio to their increase, than when the proper number of bees only occupy the hive.

A LACK OF WORKERS DISASTROUS TO THE FAMILY.

The same, or rather equally disastrous effects follow the lack of a sufficient number of bees to perform the necessary labors of the family. The queen requires some five or six combs, about twelve inches square, in which to deposit her eggs; and on taking possession of a new hive, on swarming, she requires these combs as soon as they can be constructed. If the swarm be small, these combs are not built until the season is so far past that they are of little use, and the chances are, that they are not built at all. Some four or five segments of combs, of about half the usual size, are all that are built generally. In these combs the queen finds but a small portion of the space, that she would use, as a receptacle of her eggs, if she had the necessary room; and even what space she has, under such circumstances, cannot be devoted to the young brood, since the sparseness of the laborers of her family calls for so many to be constantly abroad, that but a very few remain at home; consequently, the necessary heat to develope the brood cannot be generated, and the queen knowing this, will only deposit a few eggs in the centres of such few combs as she has, and do the best she can. In such cases, the queen may confine her laying to two or three places, where the most warmth can be generated, of about the size of a tea-cup, when, if she had a hive full of combs, and workers enough, she would cover some five or six, or more combs, twelve inches square; and produce more bees in one month in this way, than in a year, as before stated.

The only way for such families, short in numbers, to make up their complement, is to await another season, when, taking time by the forelock, they will have the usual numbers by midsummer.

In the foregoing case of a superfluous number of bees, it is not advisable to give more room, for the reason, that when we determine what is the exact size of a bee-hive, we should adhere to that size in *all* cases.

EFFECTS OF TOO SMALL AND TOO LARGE HIVES.

Various are the reasons for making all hives of the same size. If we make them *too small,* the bees are more liable to perish from the effects of an unfavorable winter, and from the ravages of the bee-moth, in consequence of the weak condition of the family. The queen, in such cases, as before stated, is curtailed of her necessary room, and not as many bees will be produced; and and whatever operates as a check to the production of larvæ, is a *fatal* error in the management of bees.

If we construct our hives *too large,* the bees will require two years to fill them; and the natural increase by swarming is much lessened, and, in some cases, entirely prevented for a series of years. Hives of this character are those made about fourteen inches in diameter, by about fifteen or eighteen inches in length. Such a size I consider to be entirely at variance with the natural requirements of the bee.

On the other hand, hives made about a foot in diameter, by six or eight inches deep, or eight or ten inches in diameter, by a foot in length, I consider equally fatal to the prosperity of the bees. Such hives do not afford the area of combs that a queen requires; and hence, she is debarred the opportunity of giving that increase, that she otherwise would. Such small families do not winter as well; as it has been thoroughly tested, that strong stocks winter better, and consume *less* honey than *weaker* ones! This may appear strange to those who are unacquainted with this subject; yet it is true, for the reason, that the bees are less exposed in strong stocks, to the various winter changes of weather, to which our climate is subject.

A few warm days in winter will put the whole of a small stock in motion; whereas, a strong one is much less affected; and when a family of bees is once aroused from their lethargy, they consume double the quantity of honey that they do when in a state of quietude. But setting this matter entirely out of the question, there are yet good reasons for having larger hives. When bees are placed in hives adapted to their natural wants, giving no excess of room, nor curtailing the use of such space, as they actually require, they then cast off their first swarms of such numbers as nature teaches them are best adapted to prove prosperous; and it matters not how large your hive may be, if a swarm be cast, which is seldom with families in large hives, it will not be in proportion to the size of the hive, but in accordance with the laws of nature, governing the bee.

Now, to come to the point, with as few words as possible, and do justice to my subject, I will say, that I have found, from many years of close application to the nature, economy, and general management of bees, that hives about *one foot square* in the clear; that is, in the inside, conform more to the natural habits and acquirements of bees, than any other size.

THE INSTINCT AND NATURE OF THE BEE UNCHANGEABLE.

There is not a solitary feature pertaining to the domestic honey-bee of the United States, that is not found just as fully developed in Siberia, Russia, China, Africa, Greenland, or in any other part of the world. Kingdoms may perish, and the giant oak may thrive amid the ruins of cities now teeming with life and gaiety, but the instinct and wisdom, and natural habits of the little bee, implanted in her censorium from the beginning of the world, will stand as immutable as the great Creator of all. Not all the art and genius of man can teach the bee one jot or tittle of knowledge beyond what God has given the impress of! Nor does she need man's wisdom. Perfect in every work, she stands forth an example for man, at least in her habits of industry. In her architecture, no man can imitate her. From her unchangeable course, that has marked her career since the creation of the world, no power on earth can cause her to deviate. The folly of man is now busy in prescribing limits, in forcing her to act contrary to her wonted nature, or rather surrounding her with useless contrivances, to force from her what nature has not bestowed upon her, in great and extraordinary labors and products of the mellifluous juices; but it is all time spent in vain. The honey-bee is capable of doing just so much, when she has wherewithal to do with; and it requires no stimulus from man to bring her to her task. All that man can do, is to give her a tenement suited to her wants, and if the fields afford honey, she will gather it.

There is no such thing as *laziness* with the bee. Far more depends upon the *bee-pasturage* and *season* than upon anything that man can do; yet we have our part to do also, and it is only by a proper attention to our duties, that the bee is protected in her labors, that result in her own prosperity, and to our own advantage.

RESULT OF THE AUTHOR'S EXPERIENCE IN LARGE HIVES.

In 1842 I had a few hives made 12 by 18 inches in the clear; that is, 12 inches wide, and 18 inches long. In speaking of the size of hives, I refer to the body of the hive for the dwelling of the bees, without any regard to what are termed *supers** for storifying. I found that it took the bees two seasons to fill my large hives; and when filled, they did not swarm at all some seasons; for the reason, that however great the quantity of bees may be in a hive, in the summer and fall, they dwindle away before spring, to a certain quantity; and thus leave a vacant space at the bottom of the hive, of some six inches or more, to be filled up with the increase of spring; while smaller hives are full, and are throwing off swarms in profusion. Here lies the philosophy of adapting the hive to the natural wants of the bee. I will illustrate this point by a supposable case.

An apiarian places a swarm of bees in a hive, say 14 inches in diameter, by two feet in length; the bees might possibly fill the hive with combs, the second year, but

* *Supers* are such hives, or boxes, as are placed above the regular hive, and receive the surplus gatherings of the bees, and may be removed at pleasure.

swarming is entirely out of the question with a family of bees in such a hive. The increase of every succeeding year would disappear before the following spring, or rather numbers equaling the increase; since all the bees existing in hives in the *spring* of the year, save the queen, were the young of the preceding summer and fall. Now, *ten* years have past, and this hive is in *precisely* the same condition that it was in *nine* years ago. Not a solitary swarm has ever issued therefrom. Ten generations of bees have existed, nine of which are past away.

We now pass to what would have been the result, if said swarm had originally been put into a hive about 12 inches square.

The second year, a swarm would have issued, without doubt, and perhaps two; but we'll say one, in order to be on the safe side, as it is not my intention to give an over-wrought picture in anything I that may discuss. We will now take the very reasonable, and low estimate, of one swarm from every stock, every season, and count up how many would be the result at the end of ten years. The second year, 2, in all; the third year, 4; the fourth year, 8; the fifth year, 16; the sixth year, 32; and so on—the *tenth* year showing FIVE HUNDRED AND TWELVE families from a single swarm!!! In this calculation, we allow no drawbacks to the prosperity of the bees, such as destruction by the bee-moth, &c.; yet the usual casualities attending the culture of bees, I contend, can be almost, if not wholly prevented by proper management. So confident am I, that 512 families of bees can, in *ten* years be produced from a *single* swarm, that I should not hesitate to enter into heavy bonds, (the uncertainty of life considered,) to produce that number, or forfeit the whole actually produced.

512 stocks of bees are worth, at least, five dollars per stock, amounting to the enormous sum of $2,560, while the *same* swarm, from which so vast a profit arises, if placed in too large a hive, at the end of ten years, is worth but the paltry sum of $5, with no increase! I leave the reader to his own reflections, on the wretched management of bees, as too generally practiced in every part of the country.

HIVES DIMINISHED IN LENGTH.

In regard to my large hives, I saw the fallacy of such dimensions, and concluded to try the experiment of cutting them off, nearly filled with bees as they were, which I performed with a common hand-saw; the manner of doing which, was as follows:—

It was in the month of April, that I performed the operation. I ought to have done it in February or March, but the idea did not occur to me, until those months had past. On a cool morning, I examined my hives, and found a vacant space of about six inches at the bottom of each hive, unoccupied by bees. I then set them, one at a time, on a table with the bottom-board up, in close contact with the hive, giving the bees no opportunity to escape. Having my saw put in prime order, and having secured the table against a support, to render it firm, I was then ready to operate.

Here allow me to say, that a man's success in almost any undertaking, depends upon his calmly surveying the whole ground, and foreseeing this or that result before he gets through; and being *fully* prepared and *commencing* aright. Had I taken a dull saw, and commenced this operation without securing my table *firmly*, I should have probably failed in my attempt; besides, by some mishap, perhaps I might have been mortally stung. These are small matters, it is true, yet in all of our operations with bees, it requires a nicety of calculation and philosophic view of the case; that we may coolly perform our task, and know what effect every move we make will have upon that insect, so tenacious of her rights.

Having marked off the part of the hive to be cut asunder, and having made niches on the corners of the same in order to set in the saw the more easily, I cut gently on one side, until I felt the saw perforate the combs. I then placed small wedges in the seam at the corners, and commenced on another side; when this side was also sawed through, I inserted wedges as before, and so on until I had completely cut the hives in two. The bees did not seem to be molested much, if any. I then took a small wire, about a yard long, and having wound the ends around sticks, to serve as handles, I then drew it gently and carefully across the combs, through the aperture made by the saw, taking especial care to have the wire sever them across the *edges*, rather than the sides; since that course would displace the position of them less, and much less disturb the bees. Having cut the combs entirely off, nothing remained to be done but to place the hive in its proper position in the apiary. I allowed it to remain fifteen minutes, to quiet the bees, and then went out and placed it in its position; and not a bee seemed to know that a change had taken place in the size of the hive, so tranquil and peaceable were they.

The time had now arrived for the bees to sally out, and I deferred the operation of another hive until the following morning. I thus continued cutting one off every morning, until all were finished.

Thus, it will be seen, that if any of my readers should have hives of a size that a portion of their length would be desirable to cut off, the manner of accomplishing it is easy.

After cutting off my large hives, I found that they contained no more bees than hives one foot square, that I possessed; and those of that size actually swarmed first, and had also swarmed the preceding season, while my *large* hives had not cast a swarm for a period of two or three years!

This result renders it conclusive to my mind, that it is folly for the apiarian to pay no regard to the proper size of hives, or rather suppose that the size has but a secondary bearing upon the prosperity of his bees. The *size*, sir, is *everything;* and until you learn this fact, and act upon it, your time is spent in vain.

SMALL HIVES NOT APPROPRIATE FOR SMALL SWARMS.

Some apiarians consider that the hive should conform

to the size of the swarm; rather than place small swarms in ordinary hives, and allow the bees to remain therein until they are filled by the natural increase of the family.

This is a great error; but, say they who defend this principle:—"A large family requires a large house, and a small family, a small house." This is true of people, but it has no bearing at all on the room suitable for a swarm of bees, if future prosperity and gain are to be taken into the account.

Let us take a rational view of this question. So far as the mere comfort and convenience of the swarm is concerned, during the first season, I admit that hives of such size as the bees can just fill with combs, during the first summer of their existence are best; but we must look *beyond* the first year, if we expect the greatest prosperity that is attainable.

In the first place, we must entirely discard the idea, that if a swarm be very large or very small, at the time of issuing from the hive, its existence and prosperity in succeeding years is thereby effected. I say *its* existence in succeeding years; but I mean the existence of succeeding generations of the same original family; because, no swarm of bees ever lived through *two* seasons.

In order to more fully illustrate this point, we will suppose that A has a very large swarm issue from one of his hives, say about *double* the usual numbers; and B has a very small swarm also issue, about *half* the usual size. A obtains a hive of double the usual size, for his swarm, and B looks about for a very small hive for his swarm, to suit the bulk of it.

The bees are hived and they go to work freely. At the end of the season, A finds that he has a fine hive of bees, with a good supply of honey for winter use, and on raising his hive, he finds that it is about three-quarters filled with combs. (If you give a large swarm *double* the usual room, it is not generally *all* occupied the first year.) B examines his little family and finds his little hive full of combs, but from the weight of it, he concludes that he'll have to feed the bees, in order to carry them through the winter, and he begins to wish that he had not been to the trouble of hiving them at all, and almost wishes that he had found them all dead; since the prospect before him, of feeding his little family, so as to enable it to safely pass the winter looks cheerless and forbidding. Well, the winter is past, and the genial warmth of the sunny month of May, arouses the bees to great activity. The medium-sized hives are throwing off swarms in profusion, and A wonders why his family in the *large* hive does not swarm! "I 'll get a *rouser* out of that hive when it *does* come," said A, one day to a neighbor. He might well say, "when it *does* come;" for, if he had known anything about the science of bee-management, he might have known that a swarm would never be thrown off from a hive of such unnatural dimensions.

A watched in vain for a swarm:—none came off, and on turning up the hive on the 10th of June, lo! he discovered that the bees had not added any new combs to those built the season before, and there was yet a large space of spare room unfilled by them. The *second, third* and *fourth* seasons passed away, and A's "*rouser*" had not made its appearance, and not a bee more could be discovered in the hive, at the end of that period, than he had at the commencement.

Now for B and his family. B expected one or two good swarms from his little 5 by 7 box, but he found the young bees produced in this hive were few, comparatively, in numbers, and when every other family on his premises had thrown off very large swarms, and some ten days beyond this period had past, a little weak, sickly-looking swarm did issue from this small hive, and B was sent for in great haste. After he had surveyed it for a moment, said he; "You can go. I'm not going to fuss with another *goose-egg* swarm, and feed it, to get it through the winter." He suffered the bees to perish on the branch where they clustered.

Year after year past, and B derived no manner of advantage from his little hive. It seldom swarmed, and when it did throw one off, it was very late in the season, and the swarms were so small, that they were seldom hived.

The result of the foregoing imaginary cases, is precisely what would be the consequences, of such a course of actual management. The swarm in A's hive could not, with all its natural increase, so fill the hive in the spring, as to be able to spare a single bee, since it is an invariable principle of the bee, to never suffer emigration, while an inch of their domicil remains unfilled with combs, and unfilled with bees. Let this remark be deeply impressed on your minds, ye who know it not, and much time and anxiety in regard to the swarming of your bees may be averted.

Had I been present when the aforesaid two swarms of A and B issued, I should have advised them as follows:—

Gentlemen, by all means, put your bees into the *regular-sized* hives. Yours, Mr. A, is now large, and perhaps you may, during this very warm weather, think that a common hive cannot possibly afford room for them; but you may depend upon it, that they will all find acccomodation therein. They *appear* to be more numerous than they really are, in consequence of the heat of the weather causing them to *extend* in clustering, in order to allow a current of air to pass through them. When hived, if you find a large portion to cluster outside the hive, do not be alarmed; the first few cool days we have, will drive every bee in, and next September you will acknowledge that what I say is right.

And you, Mr. B, do throw that 5 by 7 box into the fire, I entreat you. It always gives me a fit of the ague, to see the management of bees thus *butchered*, if I may be allowed the use of that term. Do get one of your *foot-square* boxes, and let them fill such a portion of it as they can. They will not more than quarter fill it *this* season, but, sir, *next* year, you will have as good a stock of bees as any in your apiary. You may have to feed this swarm a little in the fall; for, small swarms never do lay up much honey, but when the time comes for feeding, I will inform you how you can, for twenty-five cents, feed them enough in *one* day, to carry them safely through the winter, and then you will have a stock that will be worth something.

CHANGE FROM LARGE TO SMALL HIVES—DECEPTIVE APPEARANCES OF SWARMS, ETC.

When a bee-keeper is accustomed to use very large hives, or the hollow trunks of trees, called *gums* in some parts of the country, and in Virginia, in particular, the swarms are sometimes somewhat larger than those issuing from the proper-sized hives; but, as I before stated, if they pass certain dimensions in their hives, they seldom get any swarms at all; and when such large swarms do issue, if the weather be very warm, the bees extend so much, in order to allow the air to circulate among them, when in a cluster, that it is thought impossible to hive them in boxes 12 inches square. I have been written to on this subject, from various parts of the country, by those who have made use of hives that I have recommended, and the complaint was, that their swarms were so large that hives of my size could not afford the bees room; and in some instances, the bees deserted them. To persons thus circumstanced, I answer, that *appearances* are very deceptive sometimes in hiving a swarm of bees. A moderate-sized swarm, in a *very* warm day, appears much larger than it would on a cool day; and when a swarm enters a hive during very warm weather the bees find the atmosphere within insupportable, and a large portion of them are compelled to cluster on the outside of the hive, until the combs are so far advanced as to protect the interior of the hive to some extent. On such occasions, should the hive be raised, it might appear to be filled with a solid mass of bees, when, in reality, not half of an ordinary swarm are there. The deception is produced by the bees clustering on each side, within the hive, and then throwing a *sheet* of bees across the bottom, connected with festoons of bees from the top of the hive. In such cases, almost the whole of the interior of the hive is an open, unoccupied space. I have often witnessed this delusion, and in nine cases out of ten, bee-keepers would suppose that the hive was filled to a perfect *jam*. This case often occurs when a large body of bees cluster outside; and one would say that it was utterly impossible that the hive could afford sufficient room for the whole family; but let the weather change—let the wind veer around to the north, and let the sun be shut out by cold, damp clouds, and *presto!* what a change! Why, a person not in the secret, would say *positively*, that half of his bees had deserted their tenement! Instead of a hive full to overflowing, a snug, compact, moderate-sized swarm, is closely formed in the top of the hive, through which the white tips of a row of beautiful pearly combs appear.

Bee-keepers should, in very warm weather, be particular in *shading* the hives of new swarms fully and effectually, and in case of having swarms that appear to be hard pressed for room, an abundance of fresh air should be admitted at the bottom of the hive—even raise it on blocks, at each corner, one inch high. This proceeding will prevent the bees from clustering on the outside of the hive, and when the bees have been at work about a week, the blocks at the corners may be removed, and the hive lowered down to its proper position.

I cannot, however, say that cases may not happen in which all of my prescribed rules will avail nothing. I refer to cases where *two,* and even *three* different swarms issue at the same time, and cluster together on the same branch. In such cases, if the apiarian be not present when the bees swarm, he very reasonably concludes, that the whole mass is but *one* large swarm. Hives 12 inches square are of no use in such cases; that is, for the whole of them together, neither is any hive suitable for the whole of them. They should be divided, and the way to do that will be developed, when I come to a chapter on "*swarming.*"

DR. BEVAN'S OPINION ON THE SIZE OF HIVES.

That no portion of my readers may think that I am decidedly wrong in recommending hives so small as *one foot square,* in the clear, I here quote a few remarks of Dr. Bevan, an English writer on the honey-bee, whose work was re-published in this country some years ago, and circulated to a considerable extent.

He says: "In a former edition of this work, a preference was given to those of Keys, but subsequent information and experience induce me to recommend their diameter to be three-eighths of an inch less than his, viz: *eleven and five-eighths' inches square, by nine inches deep in the clear.*"

Here we have hives recommended more than *one-quarter* less in size than those that I recommend.

I have had several of Dr. Bevan's hives, or such as appear in his work, engraved, and I shall lay them before the reader; not that I approve of them at all, but being the nearest approximation to hives in use in the United States, and, perhaps, identical with many in use in this country, I think it expedient to comment on their qualities, in order to cover the whole area of my subject, or so much of it as is practicable.

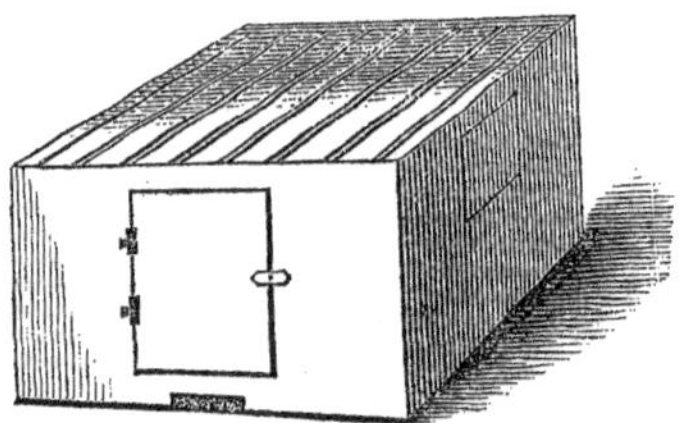

BEVAN'S CROSS-BAR HIVE.

The above engraving represents what is termed a *cross-bar hive.* The object of this kind of hive is to guide the bees in their comb-building; that combs may be more regularly constructed, thus affording more brood-combs than are generally built, when the bees are left to themselves, and less irregularities in their architecture. It is intended, that the bees shall construct their combs on the bars. The centre bars are placed suitably for brood-comb, and the outside bars are wider apart, and adapted to store-combs. This is all very well, provided the bees will follow these bars; but they will not. They must have one or two *guide-combs* attached, before they will follow the bars at all; and with this trouble on the part of the bee-keeper, not half of the time, will the bees pay the least regard to the bars, but will build combs directly across or transversely, and every other way that can be imagined.

This kind of hive is entirely too complicated for general use in this country, as well as scores of other kinds that I shall not condescend to notice. It is entirely useless to attempt to introduce into general use, any kind of hive, but such as is easily and cheaply made, and that does not require an *engineer* to put in order and oversee, as many of the *gim-cracks* of the present day do. The bee requires the simplest tenement, and many of the screws, valves, &c., now in use, are but hindrances to her natural prosperity. Let it not be understood that I would cast any imputation on Dr. Bevan, by the above remarks. The foregoing hive is not of his invention, although it was copied from his book.

I will let Dr. Bevan tell his own story relative to adjusting the bars, &c., of his hive.

He says: "The sides of the boxes should be an inch thick, and have the upper edges of the fronts and backs rabbeted out half their thickness, and half an inch deep, to receive a set of loose bars upon their tops, which should be half an inch thick, one and one-eighth of an inch wide, and seven in number. If the distances of the bars from each other be nicely adjusted, there will be inter-spaces between them of about half an inch. The precise width of the bars should be particularly attended to, and also their distances from each other; as any deviations in this respect, would throw the combs wrong, particularly if that deviation gave an excess of room. It would be better, therefore, for them to be somewhat within the rule, than to exceed it by ever so little, for whenever the bees evince a disposition to depart from the prescribed dimensions, its tendency is generally to make the combs approximate. This has induced me to have my boxes surmounted by bars varying a little in their relative distance, thus: the three centre bars are placed at the distance of only seven-sixteenths of an inch from each other, while the rest gradually recede from that distance, so that the two last inter-spaces on either side of the box, are nine-sixteenths of an inch in width. The same precision must be observed in the *length* of the bars, as it is of great importance to have them indiscriminately applicable to every box; and in case the joiner should exceed the specified dimensions of a box, the extra space must be thrown to its sides."

After these bars are adjusted, a cover is placed on the hive, of the usual thickness, and screwed down, so as to admit of being taken off at any time. Through this cover, a hole may be made some three inches square, and a super placed thereon, as in other cases.

He claims this advantage in this hive over ordinary ones; that at any time a leaf of comb may be withdrawn, and in this manner the surplus honey is obtained, or from the supers as may be desirable. Let those try this hive who choose, it is not very expensive; but I must say, that I can see nothing valuable about it.

In regard to causing the bees to build their combs with *regularity*, it is truly important to devise some method to produce such a result, and the only effectual method that can be practiced without trouble, will be given when I come to speak of my own hives, or such as were planned by me.

I condemn Dr. Bevan's hive, on account of its bars, and also on account of its size. Put three inches more on its depth and take out the bars, and it would then answer the purpose very well.

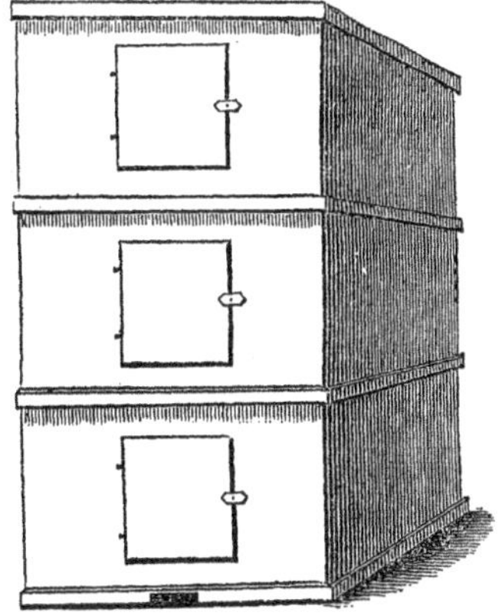

SUBTENDED HIVE.

Here is a cut of a kind of hive that is in use to some extent in this country. This is also from Bevan's work; and the size of each box is presumed to be the same as the bar-hive, viz: eleven and five-eighths inches deep, by nine inches wide. Through the two lower boxes, holes about four inches square are cut, with a slide to shut off the opening when the supers are not placed in position. The doors in front open to admit the apiarian to observe, through a pane of glass, the operations of the bees. These glass windows may be dispensed with, if one choose to do so. The opening or entrance for the bees, as seen in front, was not in the original drawing of this hive, in Bevan's work, but I have placed it there as essential, as the reader will hereafter observe. The glass windows may be in front or in the rear of the hives, according to the desire of the apiarian. If the hives be placed against a fence or wall, they should be in front; but should there be a passage-way between the hives and such fence or wall, then the doors should be on the backs of the hives, in order to observe the labors of the bees, without the least disturbance of them.

There is a hive on this principle now in use in some part of New Jersey, and perhaps in other States also, with which some *savan* is deluding the good people, by causing them to believe that it is *original*, and the very best hive in existence, of course.

HOBBY OF A PORTION OF THE ITINERANT BEE-HIVE VENDERS.

The hobby of a portion of the itinerant bee-hive venders of the United States is, "*an easy method of renewing the combs every third year.*" The idea has struck a few of those *geniuses*, that in consequence of the difficulty to the inexperienced bee-keeper attending the transfer or change of families of bees, from old to new hives, when the combs have become blackened and vitiated from several years' use, that if anything could be "got up" that would obviate the necessity of such a change, even if it ruined every other principle of correct management, money could be made by the operation, before the bauble would burst. This, of course, is a gratuitous assertion; yet I may, perhaps, be able to "look as far into a mill-stone," as any man.

TWO KINDS OF SUBTENDED HIVES.

There are *two* kinds of these "subtended" humbugs now offered to bee-keepers in the vicinity of New York, and to what extent they are used, I cannot say. One kind is on the principle of the foregoing cut, as I before stated, and the other only varies from the first, in substituting *drawers*, that slide in and out in a frame. The size of these drawers is somewhat smaller, I think, than the boxes that are placed over each other; yet the *principle* is the same.

RULES FOR MANAGEMENT IN SUBTENDED HIVES.

The rules for management in the foregoing hives, as I have it from those bee-keepers who have purchased them is, that the bees are hived in the *lower* box, and when this is filled, add a *second*, and if that be also filled, then add a *third* box. If all be filled with combs and honey, then at the proper season, the two upper boxes may be removed, and the bees expelled therefrom to return to the lower one, where the whole family should pass the winter. This is all very well in *theory*, and even in *practice*, the first and second years;

but we shall meet with this damper to our fond hopes,—an ordinary swarm will not in one case in ten, go beyond the *first* box, during the first season, if they measure about nine inches by twelve. If they be smaller than this, they will ascend to upper ones; but there is *ruin* in hives under the above-named size, in the sequel, as I think I shall fully show.

In speaking of *swarms* entering supers or boxes above the one in which they are hived during the first season, and working therein, I would observe, that in different parts of the country, the labors of bees vary according to the *bee-pasturage* about them. In a location where the white clover (*Trifolium repens*) abounds profusely, as in Herkimer county, State of New York, and some other great grazing counties, a swarm will produce much more honey and wax, than on Long Island, where the honey harvest is not so abundant.

We now return to our "subtended" hive; and we will suppose that *three* years have past, and we now wish to change our stock or family, into a *new* tenement, the old combs having existed long enough; another year, however, would not affect the prosperity of the bees, according to my experience.

Well, how is this change or transfer to be made? In the first place, you remove the box containing the bees far enough to admit of an empty one to occupy its position. You then remove the slide of the empty one, and set the *full* one *over* it. We will suppose that this operation is performed sometime during the month of April. The bees soon begin to increase rapidly, and when the original box becomes crowded, they descend and commence their labors in the lower one, having from the beginning, to pass through the lower box, to and from the hive. During the season, the lower one is filled with combs and bees, and if the hives be quite small, perhaps a third may also be filled, which may be placed on the top, or the top one may be raised, when *two* boxes only are in use, and the third placed in the centre.

October arrives, and the two upper hives may be removed, and the bees driven out, which will return to the *bottom* box, where they are to winter as before stated. The honey in the two supers removed, is the owner's gain. These supers may be removed before October, even as early as the first of August, at which time, the combs will be much whiter, and the honey better. An empty box may then be placed on the first, provided that the bees are crowded, and if any farther harvest may be expected. In the vicinity of New York, the honey-harvest is entirely past at this period, save what little the bees may gather for their daily supply.

Now we come to the grand "hobby,"—the great discovery! The bees are now in a hive with *new* combs!—just what is desired, and no trouble at all! No smoking out! No driving or whipping out! The bee-keeper is in extacies! Presently comes along the great inventor himself.—"Mr. Genius, why, how *do* you do? Let me put your horse in the stable, and you come in and stay with me, to-night. You *must* come.—John, put Mr.

Genius' horse in the stable—brush him down—water and feed him.

Mr. Genius passes the night with our extatic friend, talks over the astonishing merits of his invention, and when they part in the morning, Mr. Bee-keeper bids him farewell, adding, "you're a lucky man, your fortune's made!"

A few days subsequent to this occurrence, a gentleman passing that way, called at Mr. Bee-keeper's door to ask the favor of a glass of cool water. Mr. Bee-keeper was standing at his well, and had just raised a bucket of water. "Certainly," replied he, "water is as free as air."

"You have a fine apiary, sir,—some patent hives, I presume."

"Yes, sir, and they can't be beat."

"Pray, sir allow me to examine them; I have spent much time in studying the history and economy of the bee, and there is nothing that attracts my attention so quickly as a bee-garden."

"With pleasure, come in, and I'll show you my 'subtended' hive;—one of the greatest inventions of the age!"

"I think I have seen the same kind before. If I mistake not, every third year you can change your bees from old to new combs."

"Exactly so, sir; and here's a hive changed in that manner. Last spring the old combs of this hive were as black as your hat, and now see, (turning up the hive,) what beautiful white combs they have!"

"Just so, sir, but pardon my familiarity—there are some things connected with this change, that will sooner or later ruin your bees!"

"Ah! (looking serious,) indeed! *Ruin* the bees, do you say?"

"Yes, ruin them—destroy them—annihilate them!"

"Mercy on me! are you sure."

"Aye! positive."

"Pray, sir, what is it?"

"Look here! (turning up the hive,) do you see these thick, irregular combs."

"I do."

"You are aware that such combs are unfit for breeding?"

"For breeding?—why, yes—no, don't know as I am."

"Well, sir, not a solitary bee will ever be produced in these combs. There are one, two, three, yes, three, and perhaps four combs in this hive, that the eggs of the queen may be deposited in. They are these thin, regular combs that you perceive in the centre of the hive, which are called *brood-combs.* The others are *store-combs,* and are only made for the reception of honey. Next spring, the queen will do what she can to increase her family; but she must be restricted to three or four combs, or parts of combs, for none of them appear to be of a regular shape, as they should be; and her increase will not equal one half the number that she would produce, if she had a hive filled with the proper combs. Where is the hive that they were in last season?"

"Here it is, with the combs undisturbed."

"Now, do you perceive how regular each comb is constructed,—just so far apart, and every comb about one inch thick. Every comb here would be used by the queen, and three times as many bees would be brought into existence in this hive as in that. Here are the drone-combs on one side, a little thicker than the worker-combs. Let us examine the other hive.—Not a single *drone-comb!*"

"Well, now, I will give up. I thought that I had got a kind of hive that would be just the thing. Ah! well, it's of no use to try any of the new inventions, now-a-days. I see, sir, what you say *must* be so—I see—I see."

"Well, that is not all, sir, I'll lay a wager there is no queen in this hive."

"No queen?"

"Aye! no queen."

"What next!—John! John! (calling at the top of his lungs,) if you see old 'Genius' go past to day, tell him I want to see him. Don't let him go past, anyhow. Now, sir, be so good as to tell me—what was it? Oh! the—the queen, that's it—the queen—*No queen, did you say?*"

"Exactly so. You see that these bees are not at work bringing in pellets of farina, or what you call *bee-bread.* That hive is not so. See how busy they are! There come half a dozen with farina at once; but you see nothing of that here. The fact is, sir, that when you took off the two upper boxes, the queen was in one of them, and on being driven out of the box with the rest of the bees, she was lost, not being accustomed to going out like workers, she did not know the position of the hive where she ought to have entered. Queens are liable to be lost in this way, since they go out but once during life-time, and then they mark carefully the appearance and position of their hive. She probably entered the wrong hive and was killed by the queen belonging to it."

"*Astonishing!* What a fool I am! Are the queens always in the upper boxes?"

"Not by any means. The queen passes from one box to another, and always makes it her home where the greatest portion of brood-combs exist; consequently, she draws the most of the bees after her, if there be room for them. The hive that you just showed me filled with *brood-combs,* she was in undoubtedly."

"But they say, that if a queen is lost, it makes no difference; that the bees will make another queen."

"That is true, *if the bees have anything in the hive to make a queen from.* They want eggs or larvæ under four days old. There were both eggs and larvæ in the hive where she made her residence, without doubt; but it is very doubtful whether any were in either of the other boxes, so late as October, when you drove out the bees, and there would be no positive safety in performing that operation, even in August or September, for the reason, that there would ever be the uncertainty of having eggs or larvæ as before stated; and if they should happen to be left in the lower box, and a queen should be made, then she is to be *impregnated* by the drones, and if no drones exist, how is that to be effected."

"I see! I see! You talk like a book. *I've been humbugged, and no mistake!*"

The reader will excuse the foregoing digression from the regular train of my remarks on "subtended" hives, since an illustration of this kind is often more forcible than can be given in any other manner.

The "subtended" hive that I was speaking of, having *drawers,* operates in all its ramifications like the hive just described.

CASE IN WHICH TWO OR MORE BOXES MAY BE USED.

I have no objections to the use of two or three boxes together, provided that the lower one be about one foot square in the clear; in which case, as many boxes above as you please may be used; but *one* is as many as will be filled generally, and not even that in many parts of the country, if it contain over twenty or twenty-five pounds of honey. I hear of two hives, or boxes, each one foot square, being used in the western part of the state of New York, near Buffalo, with success. The family winters in one, and in the spring it is supered by the empty one which is generally filled full during the season, affording from forty to sixty pounds of honey.

I disapprove of transferring the family by a change of boxes, in order to place the stock in hives with new combs, as before illustrated. This course, I contend, is absolutely ruinous to the prosperity of the bees, in placing them in hives filled with combs, not at all adapted to breeding; and where the natural increase of the bees is prevented, the prosperity of the family is at an end. My method of effecting this change is by *driving* out, and it is attended by no difficulty whatever, and I consider it the only way that it can be done safely.

REMARKS ON SUPER AND NADIR* HIVING.

The reason why the combs built in a box placed *un der* or *above* the main hive, are not fit for a permanent residence of the bees, is, that the bees in ascending into a super, look upon such space in the light of a *store-room*, and the combs built in such places are almost always thick, and especially adapted to the storage of honey; being constructed in all manner of thickness and shapes. The same may be said of hives placed *under* the family to a certain extent. There is not so great a deviation from regular brood-combs, in hives placed *under*, as in those placed *over* the family; yet the deviation is enough to render such hives unfit for a permanent abode. The bees, when originally hived, are actuated by certain fixed principles in the construction of their combs,—the production of *brood-combs* always being the most prominent, since their prosperity lies wholly in the certainty of a rapid and extensive increase. But when bees are driven from their usual habitation into hives immediately connected therewith, or rather, when such extra room is afforded them, they take possession of it, and if there be a surplus population in the main hive, a portion of the bees will commence comb-building in

* A hive placed *under* the stock.

such extra space tendered them ; and, as I before stated, they will regard such room as a space for laying up their winter stores; paying but little regard to the form and thickness of their combs, and disregarding the building of *brood-combs*, in some instances entirely. This is perfectly natural, and proper that they should do so; since the idea that their home, or main tenement, is to be taken away, and they driven out, never enters their craniums. Having already constructed all the brood-combs that the queen can use, what necessity have they for more? The regular drone-cells, so important to the welfare of every family, or of its descendants, in supers and nadirs are disregarded. It is true that their store-combs are built in cells of the ordinary size of drone-cells; but they are not suited to the raising of drones, by any means. Some of said combs measuring *three* inches in diameter, while a regular drone-comb is not far from one inch thick. There may be instances in which the combs in a nadir may be built with considerable regularity; yet to trust to such for the purpose of giving bees a change of combs, once in three or four years, is a mistaken fallacy.

Again, we are subjected to the loss of the queen, as I have already shown ; and I was recently informed by a gentleman, who had practiced this method of change, on his being made acquainted with my objections to the plan, that he had no doubt that he had destroyed the queens to his hives, in the aforesaid manner, several times; but that he should never have known what the difficulty was with them, had not my remarks opened his eyes to the true state of the case.

BOX-HIVE AND SUPER.

Above I give a cut of a hive well suited for general use; and especially for the use of those bee-keepers who have not the means to construct hives, that do not come within the most economical prices. This hive is made of pine boards, one inch thick. The lower section is entirely separate from the upper one. Its dimensions are *twelve inches square*, in the clear. The top board, or cover, projects a little to render it easier to carry, when filled. A couple of sticks, about a half an inch thick, are crossed in the hive, running from the corners to each opposite corner, and put in the centre of the hive, or as near it as may be. The same may be said of every other hive here described.

Nothing remains to be done now but to make the appertures in the top for the bees to pass through into the box above, when the upper box is on. I use an inch and a quarter bit, and make *five* holes; one in the centre, and one about half way from the centre to each corner, always being sure that all the holes will come within the diameter of the super, and have some space to spare. These holes I stop with plugs made to nicely fit, and leave the ends out far enough to take hold of, and with a slight tap of the hammer, be able to remove them at pleasure. I allow them to reach through the thickness of the cover, or top of the hive, but no farther. They should be made to fit so close, that water will not pass into the hive, through the holes when plugged.

The super, or upper box, I construct of the same diameter as the lower one, but only *eight* inches deep, instead of *twelve*, the depth of the lower box. I allow the top board of this also to project a little, say an inch. The looks of the hive is much improved by this projection, and the boxes are removed from place to place, when necessary, much more easily. When I put a swarm into the lower box, I generally leave off the upper one during the first season, because here on Long Island, the bees generally have as much as they can do the first season to fill the lower one; but in many places, both boxes would be easily filled. The spring following, I unstop the holes and put on the super. As the bees increase they enter it, and by swarming time I generally find it half filled with combs, and sometimes quite filled, and the bees densely packed within it. When a swarm goes off, the super is emptied of its bees, and sometimes of its honey, as I alluded to previously. I do not find these supers to prevent swarming, unless it be second and third swarms. I generally get one good swarm, and sometimes two. I consider that one good swarm is enough, and better than more.

The foregoing hive, it will be seen, stands on a stool about 18 inches from the ground. This stool will be fully described when I speak of "bee-stands," in an especial chapter on that subject.

The reader may observe, that this hive rests on small pins or legs at the corners; giving the bees an opportunity of entering, and sallying forth on every side of the hive. This is one of the fundamental principles of my management, which I discussed in the American Agriculturist, during the years 1846, '47 and '48; together with much other matter, as some of my readers may recollect. The way in which these hives may be raised is, by driving pieces of stout wire as thick as a pipe-stem, into the corners of the hives, so as to leave just *three-eighths* of an inch of the pin projecting from the wood. The ends of the pins should be filed off smooth, or nearly so, that the weight of the hive may bear alike on all corners, and not sink any one part into the wood beyond another. These pins will support ten times the weight of the hive, without sinking into the bottom-board, if the ends be flattened.

The reason why such iron pins are recommended is, that the smaller the pin, the less liability there is of the moth-worm, that leaves the combs in the spring of the year, to find a convenient place to wind up in a cocoon, or return back to the combs, when once precipitated upon the floor-board. If wooden blocks should be used at the corners, these worms would be more apt to run up into the hive again, by the way of these blocks, than by the way of the iron pins; and I have often found the moth-worm wound up in its cocoon, in the corners made by such small blocks of wood, say half an inch long and three-eighths of an inch thick, being inserted under hives; yet such blocks may be used when it is not convenient to get the pins. Even nails or screws would do; yet I can recommend nothing that makes an imperfect job.

The reader perceives a small orifice about two inches long, and half an inch wide, in the centre of the bottom of the lower section of the hive. This opening is expressly for use in cold weather, and in the spring and summer, when the hive stands on its iron pins, or wooden blocks, this opening is closed with a tin or zinc slide, perforated with holes to admit the air into the hive, at certain seasons during the winter, when the bees are to be shut in. This opening has a greater bearing on the prosperity of the bees, than any one would imagine. I would not do without it, or a substitute, for any consideration; since, from a misapplication of the uses of this orifice, all other measures might fail to produce a prosperous state of our apiaries. The value of this opening will appear in the "winter management" of bees.

The reader may be at a loss to conceive how the bees are to be shut in, with an opening all around the hive, besides the aforesaid orifice. It is done as follows: four small holes are made in the floor-board to suit the size of the pins at the corners, in such positions that the whole four pins can at any time be lowered therein. When this is done, the only place of ingress and egress for the bees, is the small door-way, as seen in the cut, and run the slide over this, through the wire staples placed to receive it, and you have the bees imprisoned. It is important to have an opening, with a slide perforated with holes, on both sides, that is, in the *front* and in the *rear* of the hive; to admit a free circulation of air under the bees in the winter. This is another important principle of my management; but I must not digress too far; you shall hear the whole in due time.

When small blocks are used instead of pins, the bee-keeper has only to pull them out, let down his hive, close the opening in front and rear, and the bees are shut in as before.

We now come to the upper structure or *super*, and it will be perceived, that a glass window is placed on one side. This is placed in that position to save expense. It would, perhaps, be a little better to place it in the centre; yet the joiner who made hives for me, informed me, that a considerable time could be saved in placing the windows in this position, with a sliding door to run in a groove. The sliding door may be seen in the cut drawn out. They who have but a few hives to make, would not save much in this way; and I should recommend the door to be placed in the centre, and hung with very small butts. Indeed, this door may be altogether dispensed with, by those who may so choose; yet these windows are important for other purposes besides looking in to see the operations of the bees.

The foregoing remarks on the pins or supports of the hive, as well as those on the glass-windows, are applicable to every hive that I shall illustrate, except such as are *suspended,* and which do not rest upon floor-boards.

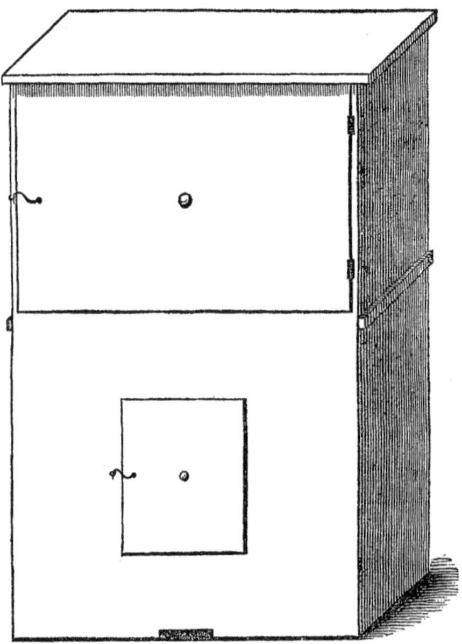

CHAMBER-HIVE.

I here give a cut of a chamber-hive adapted to the natural requirements of bees. The design and principle are not new; but I have improved on the shape and dimensions. The main body is *one foot square,* in the clear; the same size as the preceding box-hive. The chamber is *eight* inches deep, with a door hung on butts, and shutting with a small hook and staple. A glass window is shown in front, which may be omitted, if you please, as before stated. Two boxes are made of very thin boards, each with a pane of glass covering the whole front, and let into a groove in the sides cut for that purpose. There should be no *bottoms* to these boxes, but they should rest on the floor of the chamber, through which three inch and a quarter holes should be made under each box. When filled with honey, a long slender knife run under them, will easily detach such portions of the combs, as may be built down in close contact with the chamber floor or division board; and when the boxes are taken out, the bees are much easier driven out of them, than they could be, if they were enclosed on every side. If the apiarian does not sell any of his honey, it is preferable to have but *one* box to fill the whole space, because bees will work better in a single box, and lay up more honey, as a general rule, than in two small ones.

The door to the chamber, and the glass window appear in this cut to be in *front*, yet you can have either side to be the front, that you please. Both sides are adapted to be the front, or the back of the hive.

This hive is made *twenty-two* inches high, and *fourteen* inches broad. These dimensions allow one inch for the top, one inch for the division-board or chamber-floor, and two inches for the thickness of two sides—that is, one inch for each. The two sides of *full* length on either side of the chamber, are rabbeted out half an inch, so as to admit the door of the chamber to shut against the rabbet, making a better job.

The top of the hive should project all round, about an inch or more.

This hive is made to be suspended, or to set down upon a stand. There are a couple of bars, about an inch thick, placed on each side of the hive, near where the division-board separates the lower from the upper section, as may be seen in the engraving. These bars should be screwed on ; yet, for a common hive, nailing may do very well. The use of these bars is to support the hive, when the apiarian wishes to *suspend* it, rather than rest it on a floor-board, as the preceding cut represents.

This, as well as every other kind of hive that I shall illustrate, when resting on a floor-board, should rest on pins in the summer season ; and in the winter season, the bees should enter the small openings, in the front and rear only, as directed in the case of the box-hive, in the preceding cut ; and be subject to the same management in every particular.

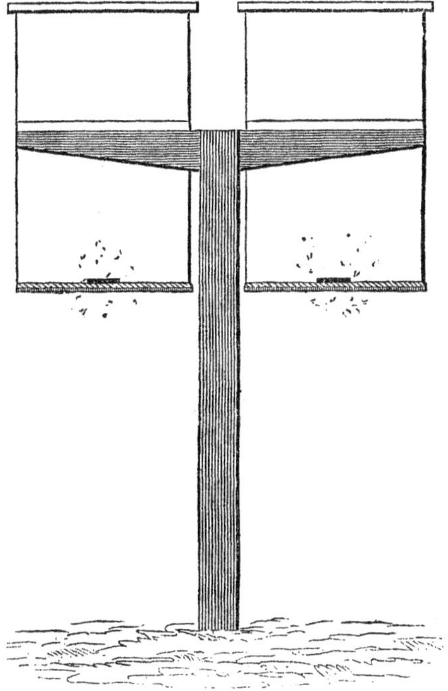

SUSPENDED CHAMBER HIVES.

The above engraving represents a couple of chamber-hives, suspended on arms nailed across joists, (timber, 3 by 4 in.) This mode of suspending hives is original; no one but myself ever adopting it, that I know of. I have also shown in the next engraving, another mode of suspending hives, of my own invention; that, for some reasons, is superior to this method.

The manner of suspending, on the above plan, is as

follows :—Take any timber, about three or four inches thick, say 3 by 3, 3 by 4, or 4 by 4, and cut off pieces six feet long ; such timber generally being about twelve feet long, one strip makes two pieces. Then sink one end in the ground, at least *two* feet, leaving the other end four feet above the ground. Then nail a strip of an inch board across the top of the post, as seen in the cut, on the side of the post towards the hives, and even with the top of it. Said cross-bar should be as small as it can be, and be strong enough to support one-half of a loaded hive, with a roof above, as will be shown. It should be broader in the centre, and taper towards the ends, as represented in the cut, in order to give greater strength. The length of this cross-bar should be about four inches longer than the width of two hives and the post ; in order to allow the hives to stand off some two inches from the post. When a post is thus set, and the cross-bar adjusted, taking care to have the bar rest horizontally, and also to have it face the exact direction that the hive should front; then you have only to set a corresponding post directly in the rear of the front one, supposing that to be the one first set into the ground, and place your cross-bar thereon, as before directed, and your stand is complete. You have, however this calculation to make, viz ; the exact distance that the posts should be set from each other, so that in sliding in the hives, a close fit may be secured. Let us suppose that our hives measure fourteen inches wide, then allowing two inches for the two bars, it follows that the posts should be sixteen inches apart. As the foregoing cut only shows the *front* view of the stand, only one post appears; the other must be imagined to stand directly behind it.

ROOF FOR SUSPENDED HIVES.

This cut represents a convenient roof for hives suspended on the foregoing plan. That every bee-stand should have protection from the scorching rays of the sun, is evident to every apiarian. I will not discuss this subject here; but will simply show how to construct a roof on the above plan, which I consider all that is necessary; or rather that this answers the purpose, with a little more attention on the part of the bee-keeper, of more costly roofing. My object is to show how these things may be done economically, as well as expensively.

According to the above cut, we take pine boards, one inch thick, and fifteen or eighteen inches wide, cut them in lengths of *four feet,* then strap two of them together, as seen in the cut; first, securing them from warping, by cleats nailed across them, on the under side, with wrought nails, and clinched. The ends of such cleats may be seen in the cut. The straps that hold the boards together at the top, may be stout leather, or butts, as the apiarian may choose. When the roof is finished, some blocks of wood may be placed on the top of each hive, in order to give a slight inclination to the sides of the roof; otherwise the two boards would rest horizontally on the hives. If the hives front the south, this roof should be drawn forward past the centre of them, and thus shade the side that needs protection, while the north side requires no shading. In the spring, when all the heat that the sun produces is beneficial, the roof may be moved back, so as to allow the sun to strike the hives, with the full force of his rays.

This kind of roof will, perhaps, require some fastening as security against very high winds, when fences and trees are prostrated. A strap, or strong cord secured to each side, directly over the posts, and then brought down and secured to the posts would be effectual. I would recommend, that an auger-hole be made both through the posts and the roof, when constructed for that purpose.

I recommend the roof to be in portions of *four* feet, for the reason, that such lengths are just sufficient for a single stand of two hives ; and such are removed with more facility than longer portions. If there are only two hives suspended, then, no longer roof can be conveniently used, nor is a longer one necessary ; but in case that a half dozen stands are existing, then longer roofs might be used, but not to advantage.

On this plan, a *single* hive cannot be suspended, since it requires two to effect an equilibrium. When more than one stand is erected, the adjoining ones should be placed at such distance, that the hives can be easily put in, and taken out, without coming in contact with the hives in the neighboring stands. If, for instance, our hives are fourteen inches wide, we should allow about sixteen inches space between the ends of the cross-bars of the different stands; thus affording facility for placing a hive in position, or removing it to another location, at pleasure.

The timber used in the posts should be equal in durability to *chestnut*, and chestnut joists 3 by 4, are the very best that can be used. It is necessary to have body to the timber *below* the ground; but *above* it is not necessary to be so strong. There is a way to economize, and at the same time beautify these posts, as follows, viz:—take a piece of 3 by 4 joist, eight feet long, and at a distance of two feet from the end, set in your saw obliquely till you come to the centre of the stick; then running the saw along through the centre four feet, you stop, and on the *opposite* side to that, on which you commenced sawing, you cut off the stick, thus giving *two* posts, six feet long, each with a shoulder two feet in length of *full* size, to be set into the ground, while the diminished portions, four feet long, above the ground, are quite strong enough, and much improved in beauty by the operation. The end of one piece will have to be squared at the top, in consequence of the necessity of cutting obliquely, to get the saw into the centre of the stick. If the joists are just twelve feet long, and by this operation the extra four feet become useless, nothing is gained on the score of economy; yet something is gained in the looks of the posts when erected. If joists can be obtained sixteen feet long, then a saving may be made, or if the apiarian chooses to lessen the height of his hives, perhaps *three* feet above, and *eighteen* inches below the ground, would answer; and in such a case posts only four feet six inches, are required, and an ordinary joist, thirteen feet long, would suffice for four posts.

It must not be supposed that this kind of bee-stand is the most beautiful that can be devised. I am now talking to the man of moderate views, who wishes a snug, plain bee-stand, at a moderate cost; yet as good as the best in practical utility.

When I come to the gentleman of leisure, with a purse ready to burst for the want of an *exit-valve,* I shall then unfold a magnificent diorama to his view; but here, among the plain every-day hives, I must stifle the utterance of these sublime views, which are reserved for those who enjoy their *otium cum dignitate.*

It may be thought by some apiarians, that stands or posts, on the foregoing plan, bring the hives too close This is not the case. The distance will be about eight inches for the two hives occupying the same stand; and the hives of the adjoining stands will be much farther off.

It is true, that bees do not thrive so well when placed in hives on a stand close together, resting on a floor-board, for the reason, that they are apt to run to and fro to each other's hives; but when the hives are suspended, this difficulty is avoided, and a bee is no more likely to enter the adjoining hive in this case, than if it were ten feet off.

There is a feature pertaining to hives suspended, not belonging to those resting on floor-boards, it is this: the alighting-boards for the bees to rest on, as they enter the hives, are in the position of an inclined plane.

Here is a *side* view of a suspended hive, with the

floor-board suspended under the hive by little wire hooks and staples. See the bees entering at the side. They enter at every side, but much more in front than elsewhere, because that part of the floor-board projects wo inches beyond the hive.

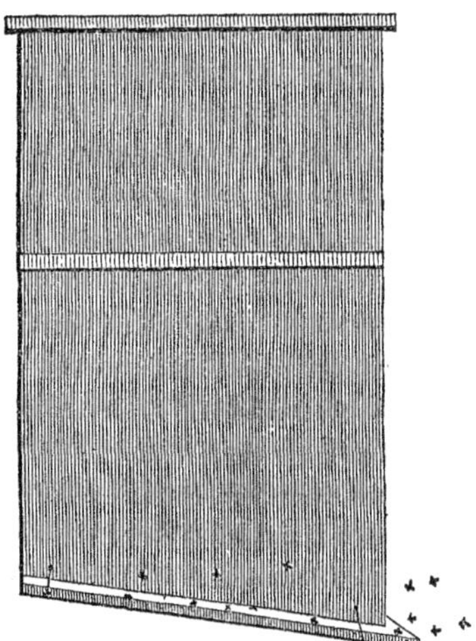

It is not absolutely necessary that the bottom-board should have an inclination from back to front in this manner; yet it is better than to have it hanging horizontally, for various reasons; one is, that it allows the water that may beat in under the hives, in storms, to easily run off; also any moisture from the interior of the hive that may drip down, readily finds its way to the ground. Again, any substances or insects that the bees have to thrust from the hive, can be expelled with much greater facility, since any one knows, that a great stone may be rolled *down hill,* that cannot be moved on level ground. The moth-worm, in the spring of the year, is dragged out of the hive much easier, with floor-boards on this plan. This inclination may be about an inch; that is, take half an inch from the *back* of the hive, and add it to the *front.* It is not best to have any projection except in front, as the suspension would thereby be attended with more trouble. The sides and back of the floor-board coming even or flush with the outer surface of the hive, the wire hooks secure it in its proper position much better than if it projected an inch or two all round.

The winter management of such hives is precisely the same, as I stated for hives resting on stationary floor-boards; so far as closing up the entrance on all sides, and compelling the bees to enter the narrow apperture in front is concerned. The method of closing the whole general entrance around the hive must be different, of course; yet the same narrow passage-way for use in cold weather, is reserved in hives suspended, as well as those not suspended. The manner of raising up the bottom-boards in the fall, when cold windy weather sets in, say in November, is by having two sets of staples, one for lowering down and the other for raising up the floor-board; or a projection of the floor-board in the *rear,* may be left, so that by sliding it forward, it will close itself; when a wooden button placed at the back of the hive, near the centre of the bottom, may be turned on its pivot and hold the floor-board firm in its closed position. The distance that a floor-board is hung from the hive is three-eighths of an inch, in all cases.

It will be observed, that in the foregoing cut of the suspended hives, no door or window appears in front. In this case, the door of the chamber is supposed to be on the back of the hive, and that of the glass window below, on the same side. If there be a passage-way back of suspended hives, it is best to have these things in the rear. For the use of the thorough, practical apiarian, the glass windows in the body, or lower section of the hive, are of little value; but for the amateur apiarian, let them be inserted, if he is willing to pay for the extra expense that it will incur, of about fifty cents each, or perhaps less.

Every hive, whether suspended or otherwise, would be benefitted by having a floor-board on the inclined-plane principle; yet it is attended with some trouble to have such, when the hives rest on a bench or stool. I have, however, obviated that difficulty, in a new hive that I have recently constructed, denominated the "Equilateral Bee-Hive," the engraving of which appears in this work. It is effected by beveling off the floor-board on every side, forming a slight cone in the centre, with the inclined sides diverging therefrom.

SECOND PLAN OF SUSPENDED HIVES.

I now introduce a second method of suspending hives, which I consider preferable to the first, on some accounts. On this plan, only *three* posts are used for suspending two hives; whereas, on the other plan, *four* are necessary. In this case, the bars that are attached to the hives to support them, are placed on the sides, instead of on the fronts and backs. Then corresponding bars are nailed across the ends of the posts, even with the tops of them, and of the same length and size of the bars on the hives. The ends of these bars are seen in the cut, as they appear when correctly adjusted. The posts, on this plan, stand opposite the centres of the hives; and the cross-bars on the posts being so placed that an equal length projects on each side. The weight of the hive, bears on the bars where they are nailed to the posts; or rather, the weight is equal on each side of the *fulcrum* or centre, and the bars are able to sustain a very great weight.

The remarks relative to posts, the hanging of the bottom-boards on an inclined plane, the construction of a roof, &c., are all applicable to these hives, as well as those hung on the first-mentioned plan.

The door to the chamber, and the glass window in the lower section of the hive, here appear in front. They can, as I before observed, be on either side; yet I think hives have a better appearance to have them in front.

It is desirable in erecting a bee-stand, to have as little shelter for insects as possible, and here lies the advantage of this stand, to some little extent, over the one first named. Everything fits very closely in the above method, affording less crevices for moth-millers, spiders, &c., than the other mode. The difficulty in the first case, merely lies in the necessity of the cross-bars being much wider than those of the second case; and as the posts will gradually work out of position, in a small degree, openings between the said bars and the hives will appear; and unless the apiarian uses a brush to clean out these crevices quite often, they become filled with spiders' webs, and various insects that do no good to the apiary; yet with care on the part of the attentive bee-keeper, there is nothing to fear. If a channel appears between the bars and the hives on the above plan a brush-broom will clear out any insects that may get a lodgment there very easily; but in the first case, there is more difficulty in effecting a dislodgment in consequence of the greater depth of the opening.

There is another method of suspending hives, quite common. It consists in setting two parallel tiers of posts in the ground, three in number, of such height as may be desired; and then nailing long strips of boards, three or four inches wide, and in length, say twelve feet, if such length is desired. The posts are so arranged, that when the boards are attached, the hives may be slid in at the ends, and rest on the bars, as in the two cases that I have adduced. The difficulty in this case is, that when half a dozen hives are thus suspended, and it becomes necessary to remove any but the two occupying the ends of the stand, it cannot be easily done, unless the bottom-boards are taken off, and the hives raised up perpendicularly, which is not convenient. Some persons may suppose that hives need not be removed at all, at any time; yet such is not the case. Hives should never be removed in the spring or summer season, unless an artificial swarm is to be made, or some operation performed that is necessary; and every thorough, practical bee-keeper will often see the necessity of removing his hives for such purposes occasionally.

TOWNLY'S HIVE.

Mr. Edward Townly, of the city of New York, has, during the last ten years, disposed of many hundreds of his "patent premium hives," as I have been informed. I have not thought it expedient to furnish an engraving of his hive, since there are so many other styles, that I consider preferable, and I shall be compelled to confine my remarks to as small a space as possible, and do justice to my subject. I understand that Mr. Townley has recently removed to the west. Whether this exchange of location is in consequence of a decline in his business I cannot say, but I infer that it is.

The business of patent-hive making has been carried to a considerable extent in this country, and by men, in a great many instances, who have had but a little practical experience in managing bees; consequently, they are not the proper persons to construct hives of the best utility, as it is out of the question, for any man to invent a bee-hive, that that shall excel all others in the long run, unless he be a thorough, practical apiarian. It is necessary to know the true reasons, why hives should be thus and so. A hive made at random that shall differ from all others, and appear to the world, who are not capable judges, to possess great merit, may not be worth a straw. Most any kind of hive, if not too large, will show a much better result the first year than subsequently; and it is in subsequent years only that their true value is demonstrated.

As to the intrinsic value and estimation in which I hold Mr. Townley's hives, it would not, perhaps, become me to declare; but I must say, that from the evidence around me of the decay of families of bees in his hives, should I speak my true sentiments, I might be accused of discourtesy. Perhaps the reader may expect to hear my objections to his hives. I will state them briefly,

The dimensions of Townly.'s hive are, for the *lower* section, where the bees have their permanent abode, about *ten* by *fourteen* inches ; and since I discussed the size of hives in the Am. Agrt., in 1846–7, I learn that he has constructed some of his hives nearer my size, about *one foot square.* The chamber, or super, projects over the main body of the hive, on every side, some three inches, being raised to admit the boxes. It turns on hinges placed on one side. The communication from the body of the hive to the boxes in the super is by holes somewhat similar to my own method. At the botttom of the hive is a screen made of wire, which is represented as affording fresh air, and at the same time, protecting the hive against the bee-moth. About an inch from the bottom of the hive a tube is inserted, about six inches long, with a bore about an inch and a half in diameter, through which the bees enter and depart. Near the top of the hive, in front, another similar tube is placed, for the ingress and egress of the family.

Now, in the first place, if he still makes his hives 10 by 14 inches, as at first, I consider that size as entirely too small. The solid contents of such a hive is much less than a hive 12 by 12 in., because the *fourteen* inches is the *depth,* not the *breadth* of the hive. In the next place, I condemn his wire screen as ruinous, rather than beneficial, to the bees; at least, doing no good at all. The only way to ventilate hives is by giving ingress and egress on every side of the hive, as I have shown in the suspended hives, before illustrated.

Again, the *upper* tube is downright ruin to **any**

family of bees. On this principle, a current of air is constantly passing up through the brood-combs, where the bees are doing all that lies in their power to get up a high degree of heat, in order to develop the larvæ. If a person were to try his best to invent something that would prove the most destructive to bees, perhaps he could not hit upon a better process than that of inserting tubes in the hives near the upper part of them. I could state many things pertaining to Mr. Townly's hive that I disapprove of; but it may be supposed that a spirit of rivalry actuates me in my remarks, and I will let it pass. What I say, is the candid conviction of my mind, without regard to my own private position in the least. I might, perhaps, better say nothing about other hives; yet my readers would not be satisfied with such a course.

WEEKS' VERMONT HIVES.

Mr. Weeks, of Vermont, has invented several hives in his day, and he has also published a small work on the honey-bee, and so, indeed, has Townly. Both of these little works are of sterling merit, so far as they go; but they are but *introductions* to the subject, and I am astonished, that gentlemen having the means of unfolding the interesting habits, economy, and management of bees, should have stopped on the very *threshhold* of their subject; but so it is, and they stand not alone. Others have done the same, and perhaps I am following them; but I think the reader will, on wading through these pages, when he comes to "*finis*," exclaim, "*enough—enough—I want no more.*"

Mr. Weeks' hive, properly denominated the "Vermont Hive," is on the same principle of my suspended hives, as illustrated at page 167. The size and shape of his hive is different, however, from mine. His bottom-board is suspended by wire hooks and staples in the same manner as I have described. He also has a chamber to his hive, in which two boxes are placed with glass fronts, on my plan; but in order to obtain a greater surface for these *supers*, or boxes, and not destroy the symmetry of the hive, he has (*as I presume*) given an inclination to the back of it. Here is a side view of one of them.

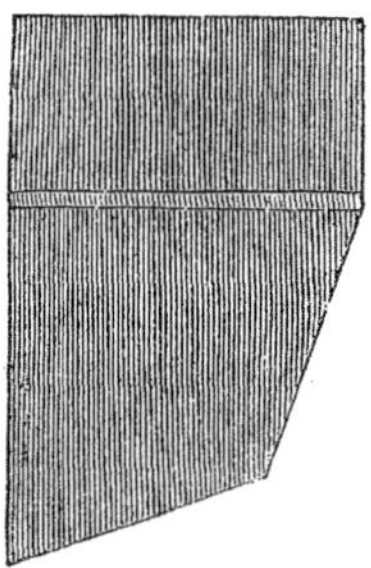

Now, this shape is not necessary at all; but if a man expects to have his hives "take" with the public, there must be a *mystery* about them,—a grand secret, and a novelty pertaining to them. Thus reason men of the present day, in a great measure; yet, after all, "honesty is the best policy."

Mr. Weeks gives as a reason for having the back of

his hive incline in the above manner, that it is expressly to hold the combs up, and also to carry off the sweatings, or drippings from the interior of the hive, which sometimes occur. This inclination is of about the same relative value as a *fifth* wheel to a wagon,—of no use whatever. If Mr. Weeks' hive has merits, it is independent of this inclination; yet anything for *novelty!* I disapprove of the shape of Mr. Weeks' hive, and think the size of the lower section too large; beyond these objections, it is very near what a hive ought to be; and it is *far* preferable to Townly's hive; indeed, if no worse hives than Weeks' "Vermont Hive," is palmed off on the public, we ought to think ourselves well off.

Mr. Weeks has also constructed another hive termed the "Non-Swarmer," which is entirely too unwieldly, and too costly for general use. We cannot afford to employ *engineers* to work our hives; and I do hope, that hive inventors will hereafter bear this in mind. Let us have something plain, simple, original, compact, and *economical*, and *then* you'll go it.

The principle on which Mr. Weeks' "Non-Swarmer" is based, is upon the principle of *collateral* hiving, or in other words, the placing of boxes at the *sides* of the main hive, instead of under, or on the top of it. He also *supers* this hive, at the same time, and thus prevents swarming.

I shall discuss the relative merits, of collateral hiving, nadiring, and supering, in a chapter devoted to that subject.

It is hardly worth my while to comment on the merits

of every hive that has had its brief existence, since I know of none, that is of any particular value.

COLTON'S HIVE.

A Mr. Colton has invented a hive, that I saw represented in the Albany Cultivator. How far this hive has been introduced, I am unable to say; but it cannot be of any real, practical utility. The principle on which it is constructed is something like this:—

The main body of the hive is of a triangular form; with one of its sides horizontal with the ground. On each side of the angle, are placed *three* boxes, in positions somewhat like the steps of stairs, each with its communication with the main hive. These boxes, which must, of course, be small, constitute the supers of the hive, and if the bees would fill all of them annually, it would be a very profitable hive; but this they will not do. I speak from a knowledge of what a family of bees ordinarily can perform, and if I should be shown a hive with double the room in the chamber, that a stock of bees can generally fill, I should condemn it as impracticable, however much to the contrary the inventor might assert.

GAYLORD AND TUCKER'S HIVE.

This is a hive invented by a gentleman residing at Poughkeepsie, or somewhere up the North River, if I mistake not. It is on the "subtended" principle, of placing boxes over each other. I have only to remark, in regard to this, as well as all other hives on this principle, that if it be intended to transfer the bees from old to new combs, in the manner as shown at page 142, then they will prove a failure, if the most of them have not "blown up" already.

There is another style of hive in use, to a considerable extent, that has no principle that is particularly at variance with my chamber-hive, represented at page 158, except the floor-board has a *double* inclination. It is done thus: the bottom of the hive is level; that is, having no inclination from back to front.

The bottom-board, or boards, are then placed with an inclination from the centre of the hive, about two or three inches from the bottom, towards each side; so that when the hive is viewed with the floor-boards in their places, two of them appear; one projecting in front some two inches, and slanting up into the hive to near its centre, from front to rear, and as I before stated, about three inches from the bottom of the hive; then, another projecting in the rear, or on the back of the hive, having the same inclination upward as the other. This description is given from a hasty examination, and I may possibly not be correct, as regard distances; but the general features of the alighting-board, I think, are as above represented. This hive is termed a "patent hive," in the section of country where I saw it in use. I infer, that some one, desirous of "raising the wind," by introducing a hive with some new "gimcrack" about it, that would look mysterious and novel, has taken the common chamber-hive, that is public property, and open for any man's use, and attached this *humbug* of a bottom-board, to make it "take" with the public!

STRAW HIVES.

Straw hives are not much used in this country; and they never would have been made in any country, but for their cheapness. The peasantry of Europe, who are not able to furnish their apiaries with wooden hives, still continue in the use of those made of straw. I consider this kind of hive as wholly unfit for the use of people who live in a land of plenty, and who are able to make wooden ones at a rate but a little dearer than those made of straw. Straw hives are only worthy of a state of abject poverty, and I hope that I shall never see one in use in this land of milk and honey, where every man can sit down to his "roast beef and plum pudding," and go to bed with his pockets jingling with "mint drops."

LOG HIVES.

Every one, I presume, has seen hives made from hollow trees, by cutting off the log of a suitable length, and then nailing a board on the opening at the top. This is a much better hive than those made of straw. These log-hives are called "gums," in some parts of the country. I recommend this kind of hive to those who wish to keep bees *without any expense whatever*. There is no principle of the habits and economy of the bee, that conflicts with log-hives; yet when boards are as cheap as they are, in those sections of the country that abound in logs, I think that boards should be used, even by the poor man, who studies economy in all his labors.

The log-hive is preferable to many patent hives now in use; and I can name several of them that I would not as soon use as the hollow log, if I were compelled to use either.

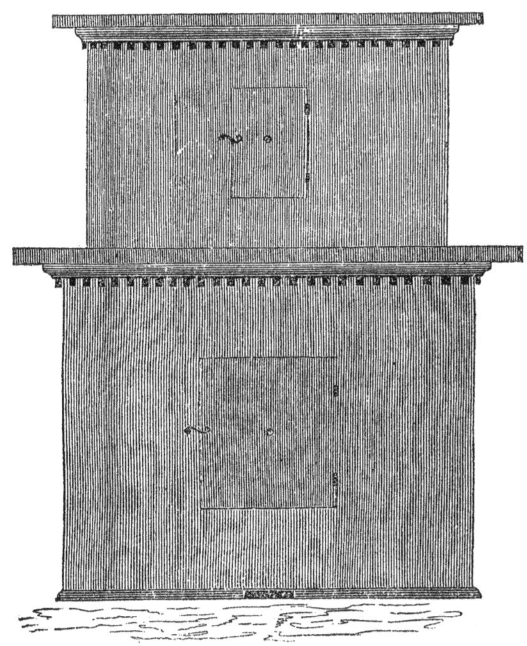

MINER'S EQUILATERAL HIVE.

The above cut represents a hive that I have constructed, with the view to combine *beauty* with *utility*. During many years of experimenting on the correct size of hives, I have demonstrated certain requisites, that every hive should possess.

Firstly, hives should be of such a size as nature will admit the bees to keep full, and yet have room enough to perform every ramification of their labors to the best advantage.

Secondly, facility to be afforded the bees in ascending to the *supers*. If we have long and narrow hives, the bees find much more difficulty in forcing their way loaded, through a long space crowded with bees, than they would through a less space. This is so reasonable, that the mere avowal of it is convincing ; consequently, we must give a more compact form to our hives, and shorten the distance to the supers as much possible, and not interfere with any other principle of management.

Thirdly, the supers should be so arranged, that the honey stored therein, may be taken with the greatest possible facility. Every apiarian is aware, that most of the hives now in use, do not offer the facility of performing this operation, that is desirable. It is true, that with a bee-dress, the removal of the boxes is not attended with any particular trouble, unless it be in the chamber-hives, where the boxes are a tight fit, and are hard to loosen from their positions ; but everything should be so arranged, that the bees will receive little or no disturbance. It is not the mere operation of removing supers, at the time that it is being performed, that we should look to. If we irritate the bees, they will not forget it for several days ; and when we do not expect it, one may plant its sting in our face ; saying, as it were, " there take *that*, for the way you jammed and knocked us about the other day."

In my EQUILATERAL HIVE, I have effected all that I think can be done, in the way of improvement in respect to the foregoing considerations. The easy manner in which the boxes in the upper section may be removed, when filled with bees, and the communication shut off with the family below, by a simple and beautiful contrivance, are very prominent features of its merits. I offer *no novelty!—no grand discovery!—no wonderful invention,* that allows the bees to produce great and unprecedented harvests of surplus honey! But I claim to have *simplified*, and divested the management of bees of its complexity, and rendered the business easy to the inexperienced apiarian.

Connected with the foregoing important results, I have beautified the general appearance of my hive, so as to render it an *ornament*, at the same time that its *utility* is admitted, and not increase the expense of making it to any amount worth taking into consideration.

The foregoing cut gives a tolerably correct view of one style of ornamenting; but I have another hive that I think surpasses this in beauty; that is, the *ornamental* portion, but the size and shape are the same as that represented by the cut. On either of these two hives, a handsome wooden *urn* may be placed, if desired, which will greatly improve their appearance.

This kind of hive may be made without any glass windows, and thereby lessen the expense somewhat, but gentlemen wishing but a few hives, should not stand for the extra trifle that it will cost to have the windows made.

The cost of making my ornamental hives is no more than the common chamber hive; and the difference in appearance is very great. Nothing could exceed the beauty of a row of these hives handsomely arranged on a well-made platform, or on stools suitably constructed, and with a tasty roof to protect them from the heat of the sun at noonday, which is all that hives require. I am strongly opposed to close bee-houses, as the reader will learn by my remarks on that point hereafter.

My equilateral hive is intended to rest on a floor-board, beveled on its sides as before alluded to, and projecting two or three inches all round the hive. The small entrance for the bees seen in front, is for use in winter, and during cold spring weather. In the summer, the hive rests on four small pinions or legs, three-eighths of an inch high, as represented in describing the box-hive at page 153. When it is necessary to lower down the hive, the legs are let into small holes in the floor-board made expressly to receive them, and very near to the position of them, when the hive is raised up as it stands during the summer.

The full particulars of every part of this hive cannot be given here, and do justice to myself. It is a true saying, that "the laborer is worthy of his hire." The production of this hive has caused me much mental labor, and I think that I am justly entitled to reap the benefit, *to a trifling extent,* that will result to the public from its adoption.

I intend to offer this hive for sale at a very moderate rate: and also to furnish *full* and *complete* drawings of *every* part thereof, to gentlemen residing at a distance, or otherwise, accompanied by a neat pamphlet, giving the most ample details in regard to every thing connected with it, or pertaining to its construction; as well as the proper management of bees in this kind of hive—all for the reasonable sum of *two dollars,* which will entitle the applicant to make as many hives as he may require for his *own* use. As this book will exist, when the author has past to "that bourne whence no traveller returns," the above remarks apply no longer than it may please God to spare his life. An advertisement will probably accompany each edition of this work, relative to furnishing the before-named hive, or the engravings, &c., which may be found at the sequel of the treatise.

COLLATERAL HIVING, ETC.

Besides supering and nadiring, there is yet another method of obtaining the surplus honey gathered, termed *collateral* hiving. This system consists in placing boxes at the sides of the hives, instead of over or under them. The following cut represents a couple of boxes on this plan.

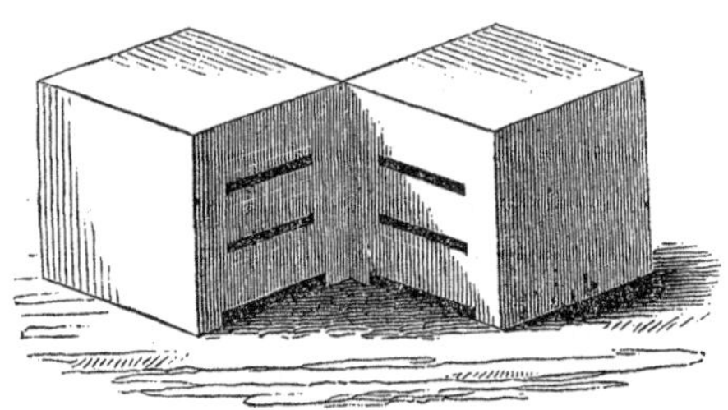

In one of the above boxes the family of bees is supposed to be permanent residents; and if success is to crown the efforts of the owner, as I view the subject, the box where the bees pass the winter, should be a foot square, or near it. Some apiarians think, that a certain number of inches in width will cause the bees to construct a certain number of combs; that is, a box twelve and a half inches square will admit of nine combs being made, whereas, one twelve inches square will only afford room for eight leaves. According to the width of brood-combs, and the interstices between, there is an abundance of space in a box one foot square, to construct nine combs; but the bees will only make eight, because the outside leaves are generally store-combs, and thicker than those built expressly to rear the larvæ in. No more than eight combs, as a general rule, would be built if the other half inch were added. I have a remedy for this difficulty, which will appear hereafter.

The English method of collateral hiving on the above plan, is to have two boxes about ten inches square, and to be put together with hinges on one side; and when closed, secure them by a hook and staple. The communication from one box to the other, is by two or three such horizontal openings as appear in the cut, besides that at the bottom. The sides of the hives that come in contact, are but half an inch thick, instead of one inch, the thickness of the other portions of them. The covers or tops, are screwed on, and the loose bars are used as represented in the cross-bar hive at page 138. When the honey is taken from the collateral box, the lid is taken off, and the leaves of combs extracted, when the bees return to the original box.

RELATIVE MERITS OF SUPERING, ETC.

The foregoing plan of obtaining the surplus honey bears no comparison to supering or placing the box *over* that occupied by the family. There is not a solitary feature pertaining to it, that recommends its adoption. The hives take up double the usual room; and the quality of the honey is inferior to that stored in supers, being subject to much more bee-bread and larvæ, and besides this, the bees will not produce as much honey and wax on this plan, as when supered or nadired.

There is no plan equal to supering, when we take everything into consideration. The queen seldom *ascends*; but she will go into collateral hives, and into those placed under her domicil, and absolutely destroy the honey with her brood, so far as a ready sale or the beauty of its appearance is concerned.

There are instances in which the bees seem to disrelish ascending into supers, even when there is no lack of numbers; and the same is the case with collateral boxes and hives placed under the family; but there is generally a good reason for their not entering and working in wax, that our eyes are closed to. In all cases of supering, one or two guide-combs should be secured to the top of the boxes, at the sides, and in the natural position. By this course, the bees are attracted to the boxes sooner than if no guide-combs were inserted. I do not wish to be understood, that guide-combs are absolutely necessary; because the bees will work in the supers, whether there be any such combs or not; provided there be a supernumerary portion of workers existing; yet they are inclined to commence their labors as I before observed, somewhat earlier with them.

COLLATERAL HIVES JOINED.

When the two boxes are closed, they present the above appearance, with the exception, that no hook in front is here shown to hold them together.

If any of my readers should feel inclined to try this system, since there is nothing like learning by experience, I would recommend that the boxes be secured together wholly by hooks and staples; say, one on top, and one on each side, at the centre. Have nothing to do with cross-bars; but when you take away the honey, separate the boxes a few inches, during 24 hours, and the most of the bees will return to the old combs; unless there be a large quantity of larvæ among the new ones.

If it be found that the bees do not desert the new combs at all, perhaps the queen may be among them, in which case, she would draw a portion, if not all of the bees in the other box after her, as soon as they might become aware of their isolation. A very good way is to cause a commotion among the tenants of both boxes, by shaking or beating the hives, when the separation takes place; and the bees will at once, from instinct, endeavor to ascertain whether the queen be safe, and among them, and the box that does not contain her will be certain to be evacuated in a great degree; and wholly, if there be no larvæ therein.

BEES REMAINING IN COLLATERAL BOXES—HOW GOT RID OF, ETC.

When a few hundred bees remain among the combs of a collateral box, they may be so frightened, as to be rendered perfectly harmless. All you have to do, is to beat the box well with a rod; and every comb may be cut out with the greatest facility; and as each comb is withdrawn, brush off the bees with some soft brush, which should be kept for the use of the apiary. An ordinary window brush, with a handle a foot or eighteen inches long, is what is wanted.

CASE IN WHICH A TRANSFER FROM OLD TO NEW COMBS MAY BE EFFECTED.

If it should happen, that the new combs in a collateral box, or even in any other, whether placed above or below, should be regularly constructed; that is, such as are used for brood-combs only, and devoid of those ill-shapen, thick store-combs, that generally occupy all extra room afforded the bees, in such a case, it would be safe to effect a transfer from old to new combs, on the "subtended" plan, which I so emphatically condemn. There is no general rule without its exceptions; and cases may occur in which a transfer may be safely made, if it be done in a manner that will not endanger the safety of the queen. The regular drone-cells would be lacking, but the bees would change, or cut down a portion of the store-cells, for the production of drones in the following spring. I do not wish to be understood as favoring this system at all, except in cases where we know that no injurious consequences can arise, which are not likely to occur.

ONLY ONE SUPER TO BE PUT ON AT A TIME.

In placing hives over each other, it is best to place but one super at a time, and when that is filled, place another box in its position, and raise the first one over it. The bees, in this case, having filled the upper one, and that box probably containing some larvæ, they will still adhere to it to some extent; and the spare room, being afforded between two bodies of bees, is sooner occupied than when placed over all, or when the first box is removed away, and the second occupying its position, with nothing above it.

BOXES IN CHAMBERS NOT LIKELY TO BE FILLED TWICE.

I would also observe, that in placing boxes in the chambers of hives of a small size, it is not advisable to place any dependence on having them filled twice or three times the same season, as some apiarians assert; because the bees manifest a dislike to commence labors anew, when they are robbed of their treasure. The safest way is to give them all the space that they can probably fill at first, and not disturb them at all, until the season of general deprivation. Bees will, however, often fill two sets of small boxes in a good season; but it is bad policy to trust to their doing so.

TIME TO TAKE AWAY SUPERS.

The question may here arise, at what time should the supers be taken off? It depends, to some extent, upon the nature of the bee-pasturage in the vicinity of the apiary; that is, the main reliance of the bees for their gatherings. If it be white clover mainly, the first of August is the proper time; or perhaps at any time during that month. If the supers be left until September or October, the combs are apt to become blackened, in consequence of the bees constantly passing over them. There are instances often, when the boxes in chambers may be removed in June; and where the boxes are found perfectly filled, and the cells sealed over, it is as well to remove them at once, and substitute others with a guide-comb, and if you get another harvest, very well; if not, no harm is done.

The honey, if left in boxes, should be covered with paper, or cloths, perfectly tight, in order to keep out insects. If the boxes be intended for market, bottoms should be made for them, and laid aside, and put on when the supers are withdrawn from the chambers, after driving out the bees.

HOW TO DRIVE BEES FROM THE BOXES, ETC.

You would, perhaps, like to know how to get the bees out of the boxes with the least trouble. In order to disturb the family as little as possible, carry your boxes to any dark place, where the bees can find their way out, by a little light being admitted near them, and in the course of the day the most of the bees will have departed, and returned home. Care must be taken not to leave the boxes where other bees will scent them out, and be attracted to them; unless you wish to divide pretty freely with them. You can, if you please, drive out the bees at once, with a rod which should be applied pretty freely to the sides of the boxes, with the open bottoms upwards. This way requires a person to be well protected by a bee-dress; but it makes the bees more irritable than the other method. In taking out the boxes, the greatest care should be observed to not crush many bees, as this arouses their anger to its greatest height.

OBSERVATORY HIVE.

Every apiarian who has leisure to study the habits and economy of the bee, should have one observatory hive; that is, a hive with only a single comb, of sufficient magnitude to afford space for the entire operations of a moderate-sized family. This hive, of course, must be very narrow, merely affording the necessary room to build one comb, and that must be *brood*-comb; and allow space on each side for the bees to labor, but not to cluster thickly.

The ordinary width of a brood-comb is about an inch, and the bees require at least three-eighths of an inch space on each side; consequently, the distance between the glass sides should be, say, *one inch and three quarters.* One and five-eighths inch wide will do very well, and perhaps just as well as to add the other eighth of an inch; yet I think the safest way would be, to make the width as I first stated; because, if the bees should be pressed into too close quarters, it would, perhaps, affect their regular labors materially.

The area of the sides of such a hive should be, at least, two feet long, and eighteen inches high; but a single comb of such large dimensions, would require a support in the centre of the hive. Here is a cut showing the form of such a hive, with cross-bars through the centre both ways, as a support to the combs.

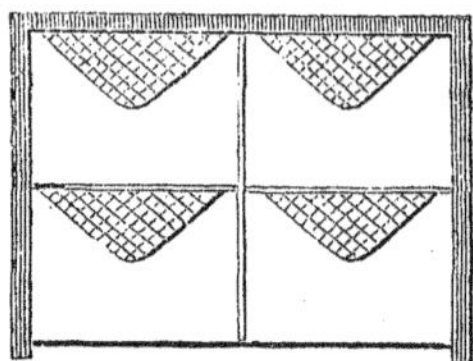

These bars should be about one inch wide and half an inch thick, supporting each other in the middle, at the junction. This size would simply occupy the same space in width that the combs will; consequently, the bees will have perfect freedom in passing over any part of the interior of the hive.

The above cut represents the comb in progress of construction in each division of the hive. The bees will often do this in this manner, when unable to work to advantage at a single point. They will even work *upwards*, when no other means affords labor to the whole of the family. Here is a cut showing the manner in which they work upwards and downwards at the same time.

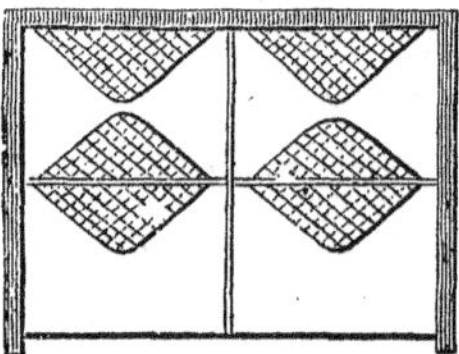

The cross-bars in these two cases afford them an opportunity of working upwards and downwards; when, if no bars were inserted, the bees would be compelled to work from the top only, since the distance from the roof to the floor, would deter them from commencing at the bottom. So perfect is the skill and architecture of this insect, that the parts of combs are united at the apex of each, with such astonishing workmanship, that it is impossible to perceive where the union takes place, or any difference from a comb worked down entirely in the usual way.

In fitting in the cross-bars, care should be taken to have, at least, *three-eighths* of an inch space between the edges of them and the glass sides of the hive; since a less space than that would, not give the bees a passage-way of sufficient diameter.

From this kind of hive, pieces of brood-comb may be easily taken, when larvæ are wanted to form artificial swarms, or for the purpose of replacing a lost queen. In order to obtain easy access to the combs, the glass sides should be hung on hinges, so as to be opened at any time, and admit the apiarian to perform any operation within, that he may choose. The glass sides or windows should be divided in the centre, and open each way, or right and left. Here is an engraving of one side of the hive, with the two glass doors closed.

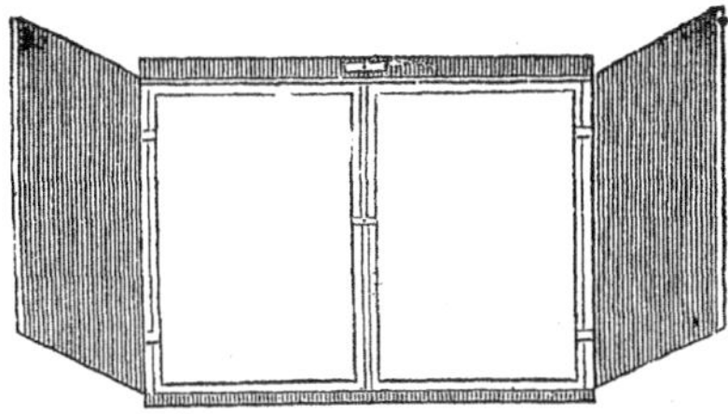

The doors should be hung with small butt-hinges on each side, being secured in their places when closed, by a wooden or brass button in the centre of the upright standard, against which, in a rabbet made for that purpose, they close. The glass doors will each contain a pane of glass about one foot wide and 18 inches long, allowing that the inside of the hive measures two feet by 18 inches, as it should measure. The frames for the doors may sink into a rabbet, planed out of the main frame of the hive, and thus admit of glass being used in them, of such size as to cover almost the entire surface of the hive. The frames for these doors should be as light as possible, and be durable and firm.

Outside of the glass doors, are to be a couple of close shutters; since the bees will not carry on their labors when exposed to the light, for any considerable length of time. The outer doors are to be hung with butts, also, and they should sink into a rabbet in the frame, exterior to that made for the inner doors. The frame for the body of the hive should be made of inch and a quarter plank, pine if you please; and every joiner can make his own calculation, relative to the proper width and thickness, to render the whole substantial, when finished. The *diameter between the two glass doors, is to be one inch and three-quarters.* This is a "fixed fact," we will suppose. The frames for said doors need not be over *half an inch* thick, and the glass can be secured in the frame, and be flush, or even with the inside thereof. The outside doors need not be over *half an inch* thick also, with clamps nailed across the ends, to keep them from warping. Now we have one and three-quarters inches to begin with, for the diameter of the inside, half an inch for each door, and being two on each side, make *three and three-quarters inches*, as the whole diameter of the frame, allowing that all the doors are sunk into rabbets equal to their several thicknesses. A joiner must be dull indeed, who cannot now make the frame-work of an observatory hive, from the foregoing illustrations.

The outside doors, when closed, may be secured in their places by a button at the top, on the frame of the hive.

In the foregoing cut, the outside doors are shown as being thrown open.

After this observatory hive is made as already defined, the question arises, how is it to be supported in its upright position? This is very easy to perform. Take a board, say two and a half feet long and eighteen inches wide; plane and smooth it nicely; nail, if you please clamps across each end, to prevent its warping; then attach it to the under side of the frame of the hive with screws, having the frame in the centre of the board, lengthwise. The board may be narrower or wider than the before-named diameter; but it should be of such width as to prevent the hive from falling over. This kind of hive should be placed entirely under cover, beyond the reach of rains and the rays of the sun, during the heat of the day.

There is yet another important consideration before we finish with this hive. We have it finished except the entrance for the bees, and that is quite necessary. The places of ingress and egress may be made by cutting out an apperture from the lower section of the frame, *under* the two doors. This passage may be six inches long and half an inch deep, on each side of the hive; thus affording the bees the facility of passing out in two directions.

The object of a hive of this character is, to witness the operations of the different classes of bees,—to see how the workers discharge their burdens—how the larvæ are fed, if *you can*—how the queen is treated by drones and workers—how she deposits her eggs—her treatment of young princesses, when sacrificed by her—her power to excite the bees to swarm, and many other interesting developments of deep interest to the scientific apiarian.

HUBER'S OBSERVATORY HIVE.

Huber constructed an observatory hive, consisting of eight frames, hung on butt-hinges, and secured by hooks and eyes when closed. There were glass windows in the outside frames only. When he wished to witness the labors of the bees in the interior of the hive, he opened the leaves as we would those of a book. The bees having become accustomed to have their hive opened in this manner, were not annoyed by the operation. In opening the leaves of such a hive, the operator must be very steady in all his movements, as sudden jars tend more to arouse a family of bees, than any other interference with them. A hive full of bees to its greatest capacity, may, at any time, be turned over carefully and set down on its top, without any protection to the operator; provided, that the hive receives no jar in the operation. The setting down of the hive on its top, must be done in so careful a manner, that the bees will not feel the force of it. Let but a slight mishap occur from inattention on the part of the apiarian, and a hundred bees will dart at his face and show him no mercy. The success of all operations with bees rests on the use of a steady hand. Not the least attention should be paid to their attacks upon you, when you are perfectly protected; and you should never attempt to do any act pertaining to them, involving the least liability of being stung, without full protection to every exposed part of your person. Running and dodging to get out of the way of bees, is but an incentive to still further attacks from them.

I have not considered it expedient to give a cut of Huber's leaf hive, for the reason, that I do not believe that any of my readers would ever attempt to construct one of the kind. It is expensive, cumbrous and useless; since all that we desire to see may be witnessed by the use of the single leaf hive, that I have described.

In the use of my leaf hive as before described, there may be some difficulty in getting a swarm to enter, provided the bee-keeper has had no experience in this business. A large swarm should never be selected for a leaf hive. The opening for the bees to enter on each side, should be much larger than those that I have discribed for other hives, to be used in winter, in order to afford the greater facility to the swarm in entering the hive. These openings may be cut on a bevel, sloping down to the board upon which the frame stands. If the apiarian choose, he may make any openings for the swarm to enter, that his own judgment may suggest; for instance, holes may be bored an inch in diameter in the end pieces of the frame, and near the floor of the hive, and when the bees are hived, they can be plugged up or left open. I should leave them open in very warm weather. If it be found that the bees will not readily enter, one door may be opened a few inches, and a cloth thrown over the hive, to extend down to within an inch or two of the bottom; then the bees will enter, and at evening when they are fully clustered within, the door may be closed. Perhaps the door may have to be closed by degrees, say partly at evening and fully in the morning, in consequence of a portion of the bees clustering along the rabbet, into which the door closes.

There are many things pertaining to the management of bees, that must ever be treated according to the best of the apiarian's judgment. Every case that may come within the scope of his experience, cannot be anticipated in any work on this subject; therefore, if any one should, at any time, find himself in a dilemma in his management of this insect, and find no especial rule in this Manual for his guidance, let him use the best of his judgment, according to the general principles here laid down. I do not think that anything of a serious nature will ever occur to any one engaged in the culture of the bee, from which I shall be accused of withholding information, that I ought to have given to the public. That I shall omit some things that would be well to insert, I have no doubt. Indeed, to write a work of this character, and not do so, would be beyond the power of man.

Here is something in point. I came very near forgetting to inform you, that before you place a swarm in your observatory hive, you should attach two or three pieces of guide-comb to the roof of the hive. Take the tips or edges of any new comb that you can obtain; say pieces two or three inches long, by an inch or more wide; cut them off evenly and smoothly, with a sharp carving-knife; and then, with the aid of a little melted bees-wax, attach them in the centre of the upper section of the frame or roof of the hive. Perhaps I may as well inform you at this place, how to melt the bees-wax in the best manner, and how to attach the comb.

In the first place, you want a little tin pan about six inches long, and three or four inches wide, and one inch deep. Place your bees-wax into this pan and melt it; then take a small brush, about as large around as a pipe-bowl and lay some of the melted wax, as quickly as possible, upon the place where your piece of comb is to be attached; and before the wax thus laid on has time to cool, you should dip that edge of the piece of comb to be secured in position, into the pan as quickly as possible, taking it out quickly to prevent its melting, and as soon as a coating of wax is obtained, then join it to that laid on the roof of the hive, taking particular care not to move the comb in the least, after its first adjustment. This whole operation must be done with a dextrous hand, while the wax is yet pliable, on the roof, as well as on the comb to be attached. The first trial will prove a failure with the amateur apiarian, I have no doubt. With old combs, the difficulty of attaching is not so great as with new combs, that are tender and brittle. New combs will melt, when put into the hot wax, very easily; and it requires considerable skill to perform the operation successfully. When the piece of comb is attached in its position, which must be in precisely the same place that the bees require it, always giving about half an inch space on either side for the bees to pass over, then it may be necessary to give it further security, since the weight of the cluster of bees will often disconnect it, when we think it perfectly firm in its attachment. The further security may be given by dipping the brush into the melted wax, and rubbing a little on at the ends of the combs, which being pressed firmly by the thumb in connection with a few of the end cells, the whole, when cooled, will afford perfect security.

The brush that I use, is a small paint-brush, but any one can make a brush with bristles or hair, to answer the purpose. When no brush is at hand, a swab made by tying a rag on the end of a stick will do in the place of something better; but here I am doing wrong to initiate the apiarian into habits of carelessness, in not having such things at hand, as he should have, in order to operate with facility and success. I condemn half-way work; and a man that feels interest enough in bees to purchase a swarm, should feel interest enough in their proper management, to have such things as are necessary, to carry that management into successful operation, when the cost and trouble of obtaining them is not of the least account.

EUROPEAN HIVES.

The majority of bee-keepers of the old world still use the common straw hive, in consequence of its cheapness, or from prejudice. I say the *majority*—this includes the cottagers, who compose a majority of those who keep bees in the old world. The hives used by the many scientific apiarians of England, France and Germany, are mostly of wood, and of every shape and size that can be imagined. The box-hives, as represented at page 141, are in use to a considerable exent—that is, the same principle; but no two bee-keepers unite on the same dimensions! Huish adheres to straw hives still, with a cover on top to be raised, and having cross-bars to his hives, as represented at page 138; he cuts out one or two leaves or combs when the bees can spare them, and in this manner takes all the surplus honey that the bees can afford. I consider this method unworthy of notice, except to show the folly of men at this late day, in thus adhering to a custom that is founded in ignorance and prejudice.

Of all the various styles of hives used in England, and on the continent, I find none that I can recommend to the bee-keeping community. There is the same desire for *experiment* and *novelty* exhibited there, that is manifested here. Occasionally a hive is brought forth as doing wonders; but a few years' experience consigns it to oblivion. The same spirit is extant there, that in our own country cries "*vive le bagatelle;*" and inventors are never at a loss to find a public to fleece of their loose cash, in exchange for hives, not worth the nails that hold them together.

POLISH HIVES.

As a matter of curiosity, I will give a brief description of the kind of hive used in Russia, Poland, and other adjacent countries. It is made of staves like a churn, being largest at the base. Its length is about two feet, and its breadth at the base about 15 inches. The staves are thick and clumsy, and the dimensions inside are not much over ordinary box-hives. The upper half of the hive is wound closely with rope; in order to protect it from the heat of the sun and from dampness. A board closes the opening at the top. In the fall of the year, the cutting out of a portion of combs takes place, according to the productiveness of the season. A stave is removed which does not extend beyond the lower coil of rope, and the cottager, with knife in hand, and smoke apparatus convenient, commences operations. When the bees come out rather furiously, a whiff of smoke drives them in again, and in this manner he takes away as much honey as he thinks can be safely spared, and have enough for winter use; and this method is considered the acme of perfection.

DIRECTIONS FOR MAKING HIVES.

Good, sound inch, pine boards, thoroughly seasoned,

are suitable for bee-hives. Some recommend inch and a quarter plank; but such are not necessary. In southern latitudes, the hives will require being better secured from the heat of the sun, than at the north; but no difference in the material for their manufacture is required. It is true, that plank will make a better hive than boards; yet, as a general rule, boards must be used, since plank do not come of a proper width in all cases; and, besides that objection, they are dearer than boards. Plank makes a heavy, clumsy hive, and they are objectionable on that point. Nothing less than boards full one inch thick, will answer; or rather, boards of a less thickness should never be used, because the different changes of heat and cold would affect the bees much more in hives made of thiner ones.

There has been some controversy in regard to the best material for the construction of hives. Some apiarians have recommended one kind and some another kind of boards for their manufacture; but after all, the grand secret of success in bee-culture lies not in the wood of which the hives are made. Dr. Smith, of Boston, an apiarian of considerable celebrity, strongly recommends *red cedar* for the especial purpose of keeping out the bee-moth. I have no doubt of red cedar being an excellent material to make hives of; and were it as plenty and as cheap as white pine lumber, I should say, use it by all means. In regard to its keeping out the moths, I do not believe any such thing. I believe, that if any wood possess an odor so offensive as to prevent a bee-moth from entering a hive, the same odor will drive away every bee also.

DIRECTIONS TO THE JOINER.

The joiner, in constructing the hives, should be very particular to have close joints; as every open joint will be filled by the bees with propolis, at a great expense of their valuable time. The nailing of the hives should be particularly attended to, as they are liable to spring open after being exposed to the weather a few months. Nothing less than *tenpenny* nails will answer the purpose; and then, some of them should be driven obliquely, or what the joiner calls *toed,* which will prevent the joints opening. The safest way, however, is to halve out, or rabbet the edges of the boards, so that when put together, they may be nailed *both ways.*

The doors to the windows should be beveled on every side, except where they are hung; and the door-way of the hive should have a corresponding bevel. This prevents open joints, and the doors not closing in damp weather. Every door should be clamped at each end, to prevent warping, and so should the floor-boards also.

Where hives are exposed to the sun a portion of the day, it requires the greatest care to keep many parts of them from warping out of their proper shape.

A thin strip may be run around the inside of the window, with a rabbet, to receive the glass. Let this strip be as thin as possible. When the glass is in its place, a brad driven in against it will keep it in its position. Don't forget the cross sticks to be placed in the hives, to run from corner to corner diagonally, and in the centre. A brad in each end will hold them fast. These sticks should be half an inch square, or more.

In making the box-hive, as shown at page 153, the super or upper section will require *dowelling*; that is, a couple of wooden pins at two of the corners, to sink into holes made in the roof of the lower section, in order to hold the super in its proper position. The pins should not be sunk into the roof over half an inch, and they should be placed at the diagonal corners.

The boxes for the chambers of hives represented at page 158, should be made of the thinest materials that can be obtained. Whitewood will do very well, but any material of the thickness of segar boxes is much better. A groove is plowed out near the front end, to receive the glass. No bottoms are required for these boxes, as I have already explained, in the description of chamber-hives. There is a difficulty arising, when the boxes are withdrawn from the chambers filled with honey, in the manner of cutting out the combs with facility. What v then wish is, to be able to sever the combs from the *top* of the box. It is quite easy to cut the ends and sides, but unless we have a knife made with a right angle, we cannot separate the attachments on the upper side, without taking an end or a side off. Now, it is necessary that every apiarian should have such a knife, with ngle, as I shall give a cut of hereafter; but not one in ten will probably ever provide one; consequently, I must give such directions in making these boxes, as to obviate, in a measure, the necessity of such an instrument.

The way to construct the boxes, is simply as follows, viz: let the back ends of them be covered by the end pieces over the ends of sides and bottoms; that is, in such a manner that they can be taken off with the greatest facility. If no directions be given on this point, the joiner will slide the ends down *between* the sides; but this is wrong; they should be on the *outside* of all, so that they can be removed easily. Every part of the boxes, except the ends, should be fastened with inch brads, but the ends should be secured with the smallest brads that will hold them in their place, and as few to be used as possible. When the honey is to be taken out of a box thus arranged, a knife is to be run down at the end and sever the combs; then take off the end, and run the knife along the top of the box horizontally, and the work is done at once: then replace the end of the box, and it is ready for use again.

PAINTING HIVES.

When your hives are made, you will wish to know what color they should be painted. Some apiarians recommend *white* as the proper color, since that color does not draw the rays of the sun; but others object to white, because it attracts the moth-miller in the night, more than darker colors. I do not think it makes a whit of difference, whether your hives are white, red, black or grey, so far as the general prosperity of the bees is concerned. We should have a durable color; one that will stand the weather well. I have used a chocolate color with good results. I make it thus:—take white lead and raw oil, with which mix Venitian red and lamp-black, to produce the color desired. The relative quantities of each can be ascertained by any person, when the same is mixed. The white lead and oil should be mixed first, then add the lamp-black to produce a *lead* color; then the Venitian red, and you have the shade desired. Raw oil stands exposure to the weather much better than boiled oil; yet if you wish to have your hives dry speedily, and if the weather be not very favorable for such a result, you can use a little litharge, or, if you please, a little boiled with the raw oil.

CHAPTER XIV.

BEE-HOUSES.

The above engraving represents an ornamental bee-house, from an original design, executed expressly for this work. It is not intended for general use, but as an ornament to gentlemen's grounds or flower gardens.

This is the first design of this nature, that has been laid before the public, to the best of my knowledge. In all the various works on the honey-bee, published in the old world, I find nothing but the ordinary bee-stands of ages past, or simple sheds of no more beauty than a pig-sty or a hen-roost. That such a structure would truly be an ornament to the flower garden, every one will admit. Why, then, should such bee-houses not be erected? The cost will not be much. Fifty dollars will suffice to cover it.

SHAPE, ETC.

It will be perceived, that the foregoing cut represents an *octangular* building; that is, one having eight angles or sides. This affords accommodation for eight hives, or one to each angle. The height should be sufficient to allow a person to walk under the lower extremity of the roof with facility, and no higher; consequently, the posts should be about seven feet long. The roof should project over beyond the posts two feet, at least, in order to shade the hives during the heat of the day. The style of architecture may vary according to the taste of the owner; yet the style of the foregoing cut is not unbecoming, by any means. Instead of having a floor, as is here represented, the posts may be inserted in the ground about two and a half feet; and the area within the posts, may be graveled, so as to have a neat and tidy appearance. The portion of the posts placed in the ground, should be left untouched, and as large as possible. These posts may either be turned, as they appear in the cut, or they may be boxed in, and made with suitable mouldings, to look very well. If they be set into the ground, they should be of some kind of durable wood; and the ends to be put below the surface, ought to be charred with fire, to prevent decay. With box-columns or posts, the style of architecture should be changed. A cornice should be run around the structure; a dental cornice, perhaps, would look well. Every builder, however, will know how to give the best effect to the general appearance of the structure. If the posts be not inserted in the ground, let the floor be laid, and ordinary joists measuring three by four inches, will do for the columns, if boxed in. In this case, it will, perhaps, require some support to prevent the structure from being blown over in a gale. Three or four posts sunk into the ground even with the floor, and made fast thereto, would be all that is necessary.

ROOF—HOW PAINTED, ETC.

The roof of this structure should be of tin, and painted a brown or stone color, or any shade that may be desired. If, however, it can be covered with shingles, let it be done. Shingles will look as well as tin, if neatly put on.

There may or may not be, a ceiling under the roof. It will look better with one, and the cost will be but a trifle.

SIZE, CIRCUMFERENCE, ETC.

The size of the house should be about twenty feet in

circumference, so as to allow full two feet between the columns. This is the smallest space that hives can occupy to advantage. The circumference of the base of the roof is much more than the foregoing dimensions, in consequence of its projection.

HEIGHTH OF HIVES—FLOOR-BOARDS, ETC.

The hives may be set from two to three and a half feet from the ground. The higher they are placed, the more they will be protected from the rays of the sun and from storms. The stand upon which they are to rest should be made of a single board in width, if possible, and bracketed on the under side, to prevent warping. In joining the floor-boards of hives, there is danger of affording cracks for the use of the moth-worm to wind up in.

The width of hives is, say about fourteen inches on the outside; and the bees require, at least, two inches space in front to alight on; and the whole width of the stand would be, according to this calculation, 16 inches, which would be its least possible diameter. There may be separate floor-boards for each hive to rest on, if the owner choose, on the bevel plan, that I have described at page 169. This would be better than to have the hives rest on a level floor, when rains beat in under them; because a level floor is apt to warp some, at best. I dislike to multiply the fixtures of a bee-stand; for the reason, that every addition furnishes some crevice, sooner or later, for insects to breed in. If separate floor-boards are furnished, let them be two inches, at least, wider on every side, than the hive, and clamped at the ends to prevent warping; then, I recommend in the place of the level floor-boards stationary in the structure, as above alluded to, to simply have a couple of string-pieces, say two inches wide, by one inch thick, placed about a foot apart, and upon these lay your bevel floor-board, strewing salt where they come in contact, plentifully. If the level floor be used, a division between each hive is necessary; that is, a board six inches broad, to be set on its edge vertically, half way between the hives. This prevents the bees running over to gossip with their neighbors, where the only welcome they get, is certain death, if they enter their neighbors' domicil!

The stand for the hives should be constructed wholly inside of the columns, resting against them. This throws the hives back, and more out of the reach of the sun. It will do the hives no harm to have the rays of the sun strike them in the morning, until about 10 o'clock; and from 3 to 7, P. M. Indeed, it is quite necessary, that the sun should shine on, or near the hives in the morning.

HIVES REPRESENTED IN CUT—OPEN BEE-HOUSES PREFERABLE, ETC.

The two hives represented in the foregoing cut, are intended to represent my EQUILATERAL hive, as shown at page 181. These hives have a beautiful appearance, and if surmounted by a wooden urn, handsomely turned, the decoration would be complete. They rest on pins or legs, as before described, during the spring and summer, and in the winter they are let down, and the openings in the front and rear are used. The general rules for the management of bees in other hives, apply to these with the same force. One great advantage in an open apiary of this nature is, that it affords the least possible facilities for insect breeding. Every part is exposed, and the broom or the brush applied once a week, thoroughly, will root out every vestige of moths, spiders, wasps, &c.

I am aware that I take new ground in advocating *open* bee-houses; yet I hope to be able to convince my readers, that the ordinary close houses, fronting the south, as they generally do, are downright ruin to the prosperity of bees. It is a mistaken idea, that bees should be kept in a warm, sunny place. There is but one season of the year, that this principle will apply with benefit to them; and that is in the spring, during the months of April and May. From June to October, they want the same temperature around their hives, that exists in the open fields—no exposure to the scorching rays of the sun, beside a close fence, that keeps off the current of air that elsewhere exists, nor to be penned up in a close bee-house, fronting the south, where the heat is sufficient to broil a steak! My remarks on the labors of bees, to ventilate their hives, when thus exposed, as given at page 83, may here be read with profit.

I will simply ask the reader, if he does not prefer laboring in the shade, when the thermometer ranges at 90°? Well, so does the bee. Watch them on an afternoon, while clustering on their tenement, when the rays of the sun are most oppressive. Do you see them remaining exposed to the sun, or do you perceive them changing their position to the shady parts? They remove to the shady sides of the hive, of course, and why is it? Because the rays of the sun are too powerful; and many bees that cluster on the outside of the hive would be at work within, but for the insupportable heat there. From these considerations, we should infer, that hives should not be exposed to the full force of the sun's rays in the summer; nor be so situated, that the air will have no circulation around them.

THE HEAT OF THE SUN DISADVANTAGEOUS IN WINTER.

Perhaps of all the innovations upon the established rules of bee-keepers, that I shall make in this work, none will be more repugnant to their views than the assertion, that bees should not be exposed to the rays of the sun in the winter. Nothing in the whole management of bees is susceptible of being more clearly established, than this fact; and though I shall not at this place discuss the question in all its bearings, yet it is necessary, that I should state, that close bee-houses with a southern exposure, should never be constructed. Where is the bee-keeper who has not witnessed the loss of his bees, when coming forth from the hives when the ground was covered with snow? Now, what is it that allures them from their tenements? It is the warm rays of a winter's sun falling on the hives, where, perhaps, the northerly winds find no entrance. The poor bees see the light penetrating their domicil, and come down to snuff the balmy breeze. They look out, and a warm sun greets them, saying, as it were, "come forth and meet me; no chill pervades the air. All is bright and glittering; and old boreas is chained to northern icy shores." They come forth. All is calm and serene around their tenement. They rise on the wing, and sweep the fields while yet warm from their abode, and suddenly the cold winds that they imagined were hushed, come whistling past. They feel a chill that benumbs them, and they endeavor to return. The glittering snow blinds their vision, and they fall to rise no more. How great the destruction of life is, in an apiary thus situated, from the above cause, every person is well aware, who has kept bees in a northern climate. If there be instances in which large numbers of bees have perished in the above manner, and yet it has made no apparent difference in the prosperity of the apiary the following season, it was because the hives were well tenanted, and could, without destruction, spare a portion of their numbers; yet every bee that thus perishes, is a loss. A hive containing two thousand bees, that loses two hundred in the above way, decreases in value 10 per cent., and in the same ratio for the loss of any number or proportion of the family.

I will now introduce the reader to a bee-house that may be enclosed when necessary, and avoid all the fatalities of close houses, as they are usually constructed.

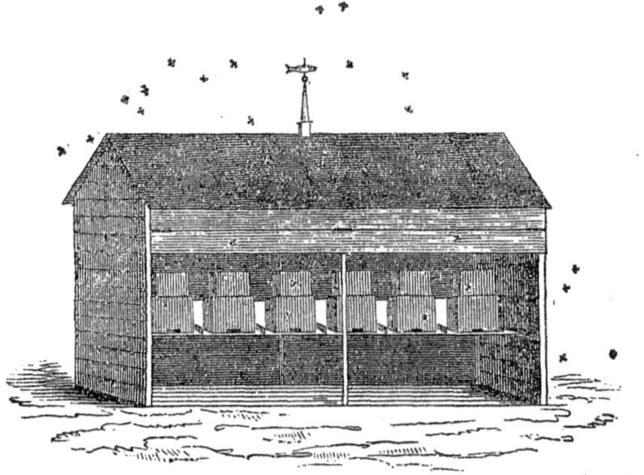

The above cut represents a house twelve feet long, six feet high, and five feet wide. The ends and back are enclosed, except a space one foot wide, directly opposite the lower section of the hives. This space is provided with a shutter, hung on hinges, and during the months of March, April and May, it should be closed. The remainder of the year, it should be open, unless in certain circumstances of very heavy winds existing, when it would be proper to close it again for brief periods. The shutter here alluded to, is made from any board measuring 12 feet long by one foot wide, and bracketed to prevent warping. During the heat of summer, a breeze will constantly be playing around the hives, when arranged on this plan, giving the bees health and activity; and during the winter, they will stay at home, where they belong.

It may be perceived in the preceding cut, that a portion of the structure is closed below the roof in front. This portion of the front thus covered, is about two feet wide. It is not intended to be permanently fast, but one foot of it, at least, in width, should swing on hinges and be susceptible of being raised and lowered at pleasure. In the spring of the year, it may be raised, and the sun let in, as the heat of this orb, at that period, is beneficial in aiding the bees to raise the temperature of the interior of the hives sufficiently to develop the brood.

HIVES TO BE BROUGHT WITHIN THE RAYS OF THE SUN AT CERTAIN SEASONS, ETC.

A very good way to bring the hives within the rays of the sun during the spring months is, to so construct the floor-board, as to admit of its being brought forward or moved back at pleasure. For instance, in March, April, and May, bring it forward parallel with the front of the house, where the sun will shine with full force upon the hives. When swarming is over and the heat becomes oppressive, let it be moved back, so far as to be beyond the reach of the rays of the sun; and in the winter, the farther back it is moved the better, for the reason, that no inducement should then be afforded to cause the bees to leave their homes, and at this season the front should be closed partially; that is, the board that hangs on hinges should be let down. The rear being open in winter, causes a cool current of air to pass around the hives, and if at any time the bees leave their domicils, they do it with their eyes open, or in other words, they are not deceived in regard to the actual temperature without, unless it be, that they find it much warmer than they anticipated, from which no evil can arise.

The removal of the floor-boards from front to rear, and *vice versâ*, will not involve the necessity of disturbing the hives. It can be effected by shoving along the whole together.

DIVISION-BOARDS NECESSARY BETWEEN HIVES, ETC.

It will be perceived, that in the cut a division appears between each hive. This is necessary, as before spoken of. A board a few inches wide, placed on its edge, is all that is requisite.

They who prefer it, may have their hives set on stools in structures of the foregoing character; and in this way, have better access to them, and facility in passing around them, &c. I am inclined to think, that setting them on stools would be the better way.

The suspended hives, before illustrated, may be enclosed in a house of this description. There is no hindrance in the least. Every apiarian must consult his own convenience and taste in many things, and not follow any written rules; or rather, he will have to do so, in the absence of instructions, since to state *every* thing pertaining to this subject, is out of the question.

COST OF BUILDING.

A bee-house on the foregoing plan, can be built for $30, and in good style, too. A handsome cornice around the roof, to suit the size of the structure, should be included in that sum. The posts should be about 4 by 4 inches, with the corners taken off an inch, except six or eight inches of the tops and bottoms. If the posts should be boxed in, they would appear much better; but for an economical house, it is not at all necessary.

FLOOR NOT NECESSARY.

A floor may, or may not be laid. If it is to give shelter to all manner of insects below, it had better be dispensed with; but if made perfectly tight, and no passage beneath be afforded, it will be an improvement. A stone or brick floor is far best, which would afford no protection to insects.

BRICK BEE-HOUSES.

Of all the bee-houses that have ever been used, none are better adapted for wintering bees, than those constructed of bricks. The great object is, to keep the bees during the winter season, in such a manner, that they will feel the sudden changes of weather as little as possible. A brick house on the plan of the foregoing wooden one, would be very convenient. An open space one foot wide on the back, would be desirable, and very important, to let a current of air pass around the hives in the summer season, at least, if not in the winter. The front may be walled up even with the floor-board of the hives; and then, a space left open eighteen inches wide; when the brick-work may commence again, supported by a cross-timber. A door-way should be left in front, to enter the building. The openings in front and rear, should be provided with shutters, that fit very closely; the one in front in particular. During the summer, the front is left open, and the hives are set back far enough, to be out of the sun the most of the day. In cold weather, the front is shut as tight as possible, door and all; and if a current of air can be made to circulate within, without the rear shutter being partially open, that may also be closed. The bees will then be in darkness, but it is so much the better for them; provided, that any means can be adopted to ventilate the apiary. A small air-hole at the bottom, at each end of the house, with an escape at the top of the roof, some six inches square, boxed in, and perforated with holes, would keep the atmosphere within perfectly pure.

On this plan, the bees will not desire to leave their hives, and the usual casualties of the winter season are entirely avoided; provided the bees have sufficient honey to carry them through the season. They will not consume over one half as much honey in this way, as they would, if exposed to the full force of the sun during the winter.

I would not wish the reader to infer, that this last method of wintering bees, is the only way that is recommendable. The preceding plan of a wooden house is similar to it, and perhaps some may think, just as good, or even preferable. The ornamental bee-house first given, is not, with all its openness, lacking qualities to enable the apiarian to winter his bees with perfect safety. A few boards so placed in front, as to exclude the sun, say a couple of posts set down temporarily, some four feet from the hives, and then boarded up six feet or more, would be all that would be necessary; then close the slides when the bees show any disposition to come out, if the ground be covered with snow, if not, let them come out as much as they please.

In case of using the brick tenement, it will be necessary to open the front occasionally, when the weather is mild and no snow exists, to allow the bees to clear their hives of dead bees, and also to their void fœces. It is very bad policy to keep bees confined a whole winter, or even a month, without giving them an airing.

The bee-houses here introduced, are original; or rather the first is entirely original in design, and the second engraving, with the plan of a brick structure, are great improvements of apologies for bee-houses heretofore existing. No apiarian has ever taken the same ground that I pursue, in regard to winter management of bees, and none have, as I believe, ever met with so successful results. I make mention of these points, not in an egotistical spirit, but rather to show that my plans are not re-vamped from any of the exploded theories of apiarians that have already existed, and been weighed in the balance and found wanting.

Had I room to spare, I would illustrate one or two more bee-houses, that might be constructed, partly ornamental, and partly otherwise; but there are none that excel those already given. Every apiarian can suggest his own plans, when the fundamental requisites are laid down, as I think I have done. The dimensions that I have given in the preceding cuts, need not be followed; but merely the principles there elucidated.

CHAPTER XV.

BEE-STANDS, ETC.

It may be necessary for me to state my views upon the relative merits of different kinds of stands, upon which to rest hives.

There is the suspending stand;—the shelf, or horizontal floor-board; and the stool-stand.

The suspended stand is a very good one, and for the purpose of giving an inclination to the alighting-board is preferable to any other; but it may be asked, how far the prosperity of the bees is affected by giving an inclination to the alighting-board?

It is not absolutely necessary to have such an inclination, yet it is an advantage in keeping the floor of the hive dry, and giving any water that may beat in facility to run off. It also aids the bees as before observed, in keeping their tenement free from worms, dead bees, &c.

The horizontal shelf has no particular fault. If it can be kept level, by the use of cleats, to prevent warping, it will do very well. The principal objection that can be brought against this kind of floor-board, is the liability of the bees to communicate with each other, when they cluster out in great numbers. When hives are set a foot or eighteen inches apart, which is the usual distance, the bees, during very warm weather, will vacate their hives, and spread out to the right and left, so as to meet the members of the adjoining families, and they frequently get so mixed, that they enter the wrong hive and perish. A bee seems to lose all knowledge of the position of its own home, except when on the wing. If they happen to cross the dividing line, between their own and a neighboring hive, they lose all recollection of having thus passed the boundary, and the nearest hive receives them; but their mistake is found out instantly, yet it is often too late to retreat. It is curious to perceive how the truant bees suffer themselves to be encircled and held prisoners. A half dozen bees will surround a single one, showing no deadly hostility, unless the stranger attempt to fly away, when it is dispatched forthwith. On an occasion of witnessing an occurrence of this nature, I stood watching the movements of a couple of workers, that held another worker prisoner. They offered no violence until the stranger attempted to rise on the wing, when it was suddenly seized by one of its captors, and stung between the rings of the abdomen. The next moment it lay quivering in death.

On refering to page 218, the reader will perceive small divisions between the hives in the cut. These strips effectually prevent bees from passing from one hive to another, as here represented. They never run up a vertical barrier, to cross over to adjoining hives, even if it be but two inches high. This being the case, a great objection is overcome to horizontal shelf-stands.

The stool-stand, as seen at page 153, is about as good as anything that can be used. It affords as few facilities for the breeding of insects as any other, and it has some features that render it preferable to either suspended or shelf-stands. It is easily removed, when necessary, and with an inclination given to each side, there can be no reasonable objection to its use. If these stools can be made in one board, they would be much better; as the groove where the joint is made, when in two pieces, will open, in time, so far as to admit the moth-worm to wind up therein. When cracks do exist, they should be filled with putty in the spring.

The size of stools should be at least two inches larger on each side, than the dimension of the hive. The clamps, to prevent warping, should not be omitted. The height may be from one to two feet. The height of hives from the ground is a matter of some importance. I have generally recommended three feet for suspended hives, and it would be better, perhaps, to have all hives as high as that, but it is not always convenient. All we want is, to get the hives out of the reach of the damp exhalations that arise from the earth during warm weather. If the hives are placed near the earth, a thick coat of gravel around them would be beneficial, in preventing exhalations of dampness. In case of using the shelf-stand, the hives can be raised three feet without inconvenience.

I do not recommend the practice of having a double tier of hives, one above the other, at all. It is bad management. The apiarian has not the facility to attend to them, that he has when but one tier exists; and besides that, it brings the bees too close.

In regard to the distance that hives should be set apart, I would say, that they cannot be placed too far, unless it be beyond the bee-keeper's premises; but it is necessary to set them near to each other, in order to afford the bees protection from the sun, &c. I think that a single row or tier of hives will not suffer injury by being placed where the space between each hive is about one foot; provided, that the divisions are put up, before alluded to. Two feet would be better, and four feet better still; but it is not always convenient to have hives that distance from each other.

The stool-stand has one advantage on this point. It can be used in an out-door apiary, and the hives stationed a rod apart, if desirable. All that is wanting, in this case, is a flat portable roof for each stool; say three boards one foot wide, and three feet long, secured together with brackets or cleats. Set one of these protectors on each hive, drawn a little forward of the centre, to produce the more shade. If they will not keep in their position, place a stone on each; but if you would be a little more tasty, you can get iron or lead weights, if any at all are necessary; which I think quite doubtful.

CHAPTER XVI.

THE APIARY.

The position of the apiary is a matter of importance. In most cases, it is seen to front the south, according to the usual practice of the present day; and especially when enclosed, somewhat on the plan of bee-houses illustrated at page 218. This position is considered necessary by bee-keepers generally, in order to afford the bees all the warmth, both in summer and winter, that it is possible to give, and which I consider so ruinous to their prosperity.

SOUTH-EAST THE BEST POINT TO FRONT.

It will not, in all cases, be found convenient to have the apiary front any point of the compass; in consequence of the situation of the ground where it is to be erected, since it is often necessary to build parallel to some fence already constructed; but the best possible way it can point is *south-east.* Directly to the south or to the east, is not particularly objectionable, when the back of the building has an opening to admit a current of air among the hives, as I have directed; but when it is convenient, I recommend a preference to be given to the aforesaid direction.

MORNING SUN NECESSARY.

Every husbandman knows full well how much more labor his hired men can perform, when they get to work at the rising of the sun, than when they lie in bed until that luminary peers in at the windows of their bed-rooms at an angle of 20° or 30°. To the above may be likened the sallying forth of the honey-bee. It is not often that bees sally forth to the fields in the morning, until the rays of the sun strike their hives. For example, two hives may be placed in the months of June, July and August, in different situations; the one where the sun cannot shine upon it, until 7 o'clock, A. M., and the other, where his rays will fall upon it, at half-past 5. Now mark the result. The bees in the hive where the warmth of the sun reaches them at half-past 5, will be seen leaving their hives at that hour, while those of the other hive, remain within until 7 o'clock, one hour and a half later. Thus it may be seen, that it is important to so place our hives, that they will receive the morning sun. If the bee-house front the south, it would be well to have a movable shutter at the east end, to be raised during the summer; say two feet space opposite the end hive, to be thus open, and closed at pleasure.

OFFENSIVE SMELLS DETRIMENTAL.

It is advisable to place the apiary out of the reach of

nauseous and offensive smells; and not immediately in the vicinity of the barn-yard, where flies congregate. A yard-house being near, will not, in ordinary cases, be injurious, unless it be offensive, which should never be the case, on account of the apiarian's family. A barrel of lime or plaster thrown into the sink, when offensive, will thoroughly purify it.

THE SHADE OF LARGE TREES NOT BENEFICIAL, ETC.

It is not advisable to place an apiary under trees of magnitude; since the drippings therefrom, during wet weather, continue long after the sun appears, and thus retard the labors of the bees. Hives that are set in an out-door apiary, without any protection, are much more affected by the drippings of trees than those placed in a bee-house. It is a custom with many people, to thus place their hives in the shade, in order to screen them from the heat of the sun during the summer; but it is bad management. A cover, three or four feet square, that may be made at a cost of one shilling, is much better than the protection of trees, and such cover may be removed in April and May, and the sun left to shine upon the hives with his full force, to aid in developing the brood. In case of removing the cover thus described, it would be well to use a small cover the size of the top of the hive, merely to prevent any warping or cracking of the top. I never expose the tops of my hives to the sun, as it is almost impossible to prevent such a result, sooner or later, to some extent.

DANGER OF HIVES BLOWING OVER.

In out-door apiaries, there is some danger of hives being blown over, during the prevalence of very high winds, unless secured in some manner. If the hives are set against a close fence, there is no danger from winds, unless it be a hurricane; yet I do not approve of placing them against fences or buildings at all, for reasons before given, in regard to allowing the air to circulate freely around them, and keeping them free from insects. It is always best, in cases of out-door apiaries, to fully secure the hives in some way, against any possible contingency. A thunder-storm in summer often brings winds that level trees, fences, and even houses with the ground. On that account, the lower the hives are placed the better; but nearer than one foot to the ground will not answer at all, and the higher the better, so far as the bees are concerned; but no height will prevent the moth-miller from entering. What fastening or security for hives is best to prevent them from being blown over, I hardly know, but a stake driven firmly into the ground, against the back of each, and a leather strap or cord running around the lower section, and secured to the stake would be effectual. The super, if the hive be on the plan of that shown at page 153, will not be blown off, if doweled in; and even if it were not doweled, it would not blow off, as the bees always cement down supers with propolis, so as to require a considerable force to separate them from the main hives.

The cheapest way of holding down hives, is to place

a large stone upon each; so let no bee-keeper suffer his hives to be blown over for the want of means to secure them.

SURROUNDING PROTECTION NECESSARY.

When bees arrive within a few feet of their hives, it is very important that the force of the winds should be checked in some manner, as the greatest difficulty a bee encounters when on the wing, is to alight safely at her own door. Like a ship at sea in a gale, all goes on merrily so long as she has sea room; but let her approach land, and then comes the real danger. Just so it is with the little bee, when the high winds sweep over the hills and valleys. She beats up against the breeze fearing nothing while she has space to dart over the forest; but when she comes to her door—when she slackens her speed, she is at the mercy of every fitful gust that plays around the hives; and often when just reaching her own domicil, as she hovers slowly before the entrance, loaded with treasures from nature's storehouse, she is forced to the ground, or perchance against some neighboring hive. The vision of the bee is obscured when she approaches within a few feet of her hive, and her motion is necessarily quite slow on such occasions, and if she be driven out of her course to the least extent, she has to rise again on the wing, and describe a circle in her flight, some ten or twenty feet above the apiary, before she can venture to return again. Even should she be driven but a single foot from the point she aimed at in alighting, she would rise again, and make a second attempt in finding her home. Bees seem to know nothing at all of the position of their hive, unless when descending on a return from the fields, or in cases when their flight is merely sporting immediately around it.

For the purpose of affording a check to the force of the winds in out-door apiaries, immediately around the hives, in unsheltered situations, I would recommend a close fence to be placed some short distance from them, on the north and west sides. If any fence be placed on the east side, it should not obstruct the rays of the sun to the most easterly hives. It is advisable to have an open length one foot wide opposite the hives, that may be opened and closed at pleasure. In the spring the whole fence may be closed, and as the heat of summer approaches, the doors or shutters may be thrown open. When a quiet nook already exists, where the force of the wind is partially broken, the hives may be placed there without further trouble. When the hives are placed in a bee-house, no protection from winds is required, except to keep the back closed when the winds are very high, and it is evident that it would be beneficial to do so. In case of having an open house, like that described at page 210, some little screen afforded as protection against the winds, such as adjoining high shrubbery, or some fence within ten or fifteen feet, on the north or west side, would be sufficient; yet without any protection at all, in any case, the bees will thrive and do well; but it is better to thus afford a little protection from the force of the winds when convenient to do so.

RIVERS AND LAKES DETRIMENTAL.

If the apiarian reside on the banks of a large river, lake, or very near the ocean, he should place his apiary as far from the water as possible; as the bees are liable to be forced down and drowned when returning heavily laden. Such results occur when the bees cross the water. If the bee-pasturage be abundant back of such river or lake, they will seldom venture across, where the distance is half a mile and over; yet there are instances where bees have been known to pass several miles over water to obtain honey. I should not suffer a close proximity to the water, in any case, to deter me from keeping bees; yet what I would inculcate most deeply is, that an apiary immediately on the banks of a river, or of any other body of water, where but a few feet intervene between the hives and the water, is objectionable. Two hundred feet from the bank is a safe distance.

HOW SITUATED IN REGARD TO THE DWELLING.

It is an important consideration to so place the apiary, that during the swarming season, the swarms will be readily observed, as the bees will not always await their owner's motion to hive them. If it be convenient to place it where the servants about the kitchen, in their running in and out, would be likely to observe the bees in such instances, it would be the best position, perhaps. When bees swarm, the noise created by them may be heard many rods, where but few hives exist; but when a dozen or more strong families are in a single apiary, the usual *hum* drowns the extra noise of swarming. Any gentleman keeping this insect, should for the space of about three weeks, charge his gardener to be on the look out. The season when swarms are to be most ex pected, is from the 20th of May to the 10th of June, and during this period, if the apiary be not near the kitchen door where the servants will notice swarms, a little attention on the part of the gardener is quite sufficient.

NO WALLS OR BUILDINGS TO IMPEDE THE FLIGHT OF BEES.

When bees sally out to the fields, they depart at an angle of about forty degrees with the plane of the horizen; and no wall, or other obstruction, should impede their free passage at such an angle. It matters not what obstruction may be in the rear of the hives, provided, that no barrier exist in front.

VALLEYS MOST SUITABLE FOR APIARIES.

If one were to have his choice of just such a location as he might elect, he should select a broad valley, with gently-sloping sides, extending a mile or more. The sides of such valley should be composed of rich meadows and pasture lands; and as little as may be under the plow. Here and there should a tract of woodland intervene, and ample orchards dot the landscape; and above all, the white clover should be seen spreading its snow-white mantle in wild exuberance and profusion, beneath the feet of the herds that rove over the fertile fields. this would be a paradise for bees; yet such a paradise exists in thousands and thousands of places, where *barrels* of honey might be gathered, for the *pounds* that are now produced.

In valleys, bees have less high winds to encounter; and when loaded and returning home, it is easier for them to descend than to ascend. This requires no proof. Let the reader, when fatigued, have a mile to walk, would you prefer to have it up hill or down?

WEEDS AROUND HIVES TO BE EXTIRPATED.

It is customary with many bee-keepers to place their hives where the grass grows in the greatest profusion. This is not good policy; yet, perhaps, it is full as well, to place hives over a green sward, where the grass can be cut at intervals, as to place them where the sod has been turned over, and then allow a profusion of weeds to spring up around them. Both cases are bad management for the thorough apiarian. I now allude to out-door apiaries only, of course.

The better way is, to first throw aside the top soil, where an out-door apiary is to be situated, then throw out a foot or two in depth of the yellow barren sub-soil. In the pit thus excavated, place the top soil first removed, and let the barren sub-soil remain on the surface of the ground. Over this spread a few inches of gravel, and you will soon have a hard foundation for your hives, where but few, if any, weeds will spring up, and where every unfortunate bee that falls to the ground exhausted with fatigue, as does often happen, may rest her weary limbs in ease and safety, and finally rise on the wing, and regain her tenement.

It may not be generally known, even among bee-keepers, that when a bee falls through fatigue, that it is difficult for her to rise from a level surface. Some little headway must be secured: and for this purpose, a narrow strip of board laid on the ground, in front of the hives, will afford the requisite facility. The bees will ascend the sides of such piece of board, and from thence take a flight, that could not possibly be effected from a level surface.

Besides the above considerations, we lose a large number out of every family of bees, where weeds and grass grow spontaneously around the hives, that would not be lost in other circumstances. How many spiders lay in ambush among the weeds and grass, to weave their silken webs around every fallen bee, no one can tell, who has not carefully investigated this subject. A few bees ensnared every day is of no account, perhaps the careless apiarian would say; but let us see what figures say, that cannot lie. Suppose ten bees are thus lost daily on an average, from the 1st of May to the 1st of November. We have 184 days, and the number of bees lost, that might, with good management, have been saved, is 1840! This number would make a very respectable swarm. Is it any wonder that people do not succeed, in many cases, in the culture of bees, when they thus let them take care of themselves?

APIARIES IN THE ROOMS OF DWELLINGS.

It is a practice with some people, to have their hives placed in an upper room of their dwelling, with tubes or other channels for the bees to obtain egress and ingress. This plan may answer very well in large towns, where no yard-room exists; or in cases of having a hive or two, kept more as a source of amusement than of profit; but in no case can bees be brought into one's dwelling, or any out-house not built expressly for them, and prove prosperous in the long run. They will thrive a short period, in spite of all the disadvantages under which they labor, and finally, they are "*non est inventus*," as the constable says, when he returns his writ unexecuted. As for preventing the ravages of the bee-moth, by attempting to get up out of her reach, you might as well attempt to get out of the reach of the fowls of the air, by ascending heavenward. The practice of thus confining bees in the rooms of dwellings, is highly injurious on account of not affording a plenteous infusion of pure air in, and around the hives, which is so vitally essential to all animate nature. Think not, reader, because a bee is but a small insect, that she needs not the necessity of breathing heaven's pure ether, like unto man. Though "man is fearfully and wonderfully made;" yet the same Architect that formed man, also formed the bee, and with the same master-hand. Let not frail mortals usurp a high distinction in the wonderful mechanism of their frame, over that of a little bee; for, we find no less to excite our amazement in the one, than we do in the other. The air that man breathes, was not made for him alone, and if we place the bee where it is not found in its purity, we do her wrong, since nature never thus destined her to be.

There is another method of placing bees in rooms. I allude to allowing them to occupy a small room at large, without being subject to hives at all; I do not approve of this method. In the first place, we get no increase from our bees. They will not multiply and fill a whole room, as some persons may imagine. The natural inherent hostility of queens towards rivals, prevents such a result. Two queens cannot exist in the same family. One must be mistress of "all she surveys." It matters not how far you "extend the area of freedom," if a second queen exist, she will be found by the legitimate sovereign, and one of the two must perish, and that quickly. Should bees form detached settlements in different parts of the same room, perhaps several families might exist for a few years; but it is folly to manage bees in this way. The surplus honey is not as easily taken away on this plan, as it is when stored in supers; and all the casualties attending the prosperity of bees, from the ravages of the moth, are subject to result from this mode of management, as well as from any other method.

BEES THRIVE IN LARGE TOWNS, ETC.

Bees will thrive in a large town with a fertile surrounding country, as well as in any other place, unless it be in a situation of peculiar merit, such as in a rich valley, with bee-pasturage in its greatest profusion existing in every direction.

In every town with a population of from 5,000 to 10,000 inhabitants; and even in cities with from 15,000 to 50,000 people, bees may be kept with the best results.

Bees fly from one to two miles with the greatest facility, to obtain honey, when it cannot be obtained within that space; consequently, an apiary situated in the centre of a town, with a radius of half or three-quarters of a mile each way, before reaching the open country, would be prosperous. In the spring of the year, the blossoming trees of every country town whether large or small, afford a rich harvest of honey.

In the city of New York, hemmed in as it is, by two large rivers, I cannot say that I think bees would thrive unless they be fed. I am aware that Mr. Townly has endeavored to inculcate a different belief, and it was his interest to do so; but I shall not transcend the limits of truth, for any gain that might accrue to me by so doing.

In the city of Brooklyn, bees would do very well. There they would have a range in the interior, without crossing the river. In almost any other city in the United States, except New York, bees may be kept with profit; but not as profitably as in locations out of town. In any situation where the most of the ground is under a state of cultivation, that is, plowed up yearly, bees do not thrive, as they do where there are extensive grazing lands; but they do well in almost any place. I know of no location in the United States where they would not prosper, except in the heart of the city of New York.

CHAPTER XVII.

PASTURAGE.

THE success attending the keeping of bees depends, in a great measure, upon the character of the pasturage in the vicinity of the apiary.

Of all the resources of bees, nothing can equal the white, or Dutch clover, that abounds to a greater or less extent, throughout the whole country; I may almost say, that without the existence of this flower, it would be useless to attempt to establish an apiary; yet there is no section of the country where it does not exist; consequently, there is nothing to fear on that point. In any place where this clover is found growing in spontaneous profusion, there will bees thrive beyond a doubt. It blooms in the latter part of May, and continues in blossom to some extent, all summer; but the height of the honey-harvest from it, is during the month of June. It is from white clover that the purest and most delicious honey is procured. No other pasturage can compare with this, so far as the purity and flavor of the honey is concerned.

Next to the above clover, stands the various blossoming trees of orchards and gardens, that are spread over every fertile landscape. In the spring, the cherry, peach and nectarine trees, first invite the bee; then the apple and pear trees spread their flowery canopies over the green fields, and afford a short but rich harvest of honey.

But first of all in the catalogue of sources, whence the bee derives a spring supply of honey, is the *willow.* When all nature wears a sombre hue, with scarcely a flower upon her bosom, the willow sends forth its tiny shoots, from which the bee obtains her first gatherings. Let one but pass beneath some stately willow at this period, and his ears will be greeted as with the music of some sweet-toned æolian harp, that seems hid among its branches; but let him cast his eye above, and there a cloud of bees may be seen flying to and fro, chanting a merry song, as they lightly dance from shoot to shoot,

> Primeval bliss, without alloy,
> Where cares can ne'er their peace destroy.

Among the earliest resources of the bee, besides the willow, are the osier, the poplar, the sycamore, the plane, the snow-drop, the crocus, white alyssum, laurustinus, &c. To these may be added, the gooseberry, raspberry, and currant bushes, with sweet marjoram, winter savory and peppermint.

Alder buds and flowers afford honey during several months. The flowers of the bean, cucumbers, squashes, pumpkins and melons of all kinds, afford a large supply of pollen.

To the above may be added the sunflower, the dandylion, the hollyhock, and Spanish broom; but above all, as a source of pollen, is the sunflower. In its golden heads, may constantly be seen the industrious workers, covered with the yellow farina of this flower, and busily engaged in kneading it upon the cavities of their legs. Every bee-keeper should plant a few dozen seeds of this flower around the border of his garden, or among his potatoes. Should an occasional seed be dropped in the potato field, when planting that vegetable, say at every sixth hill, the crop of sunflowers would be valuable for the seed to feed poultry, and be of great advantage to the bees, and not lessen the potato crop in the least.

The blossoms of mustard, turnips, and cabbage, the privet, the holly, phillyrea, bramble, sweet fennel, nasturtiums, asparagus, crowfoot, dead nettle, vegetable marrow, white lily, coltsfoot, borage, viper's bugloss, mignonette, lemon thyme, teasel, furze, heath, sainfoin, &c., are much frequented by bees.

Among the forest resources of the bee in this country, the most conspicuous are the *basswood* and *maple.* From the basswood in particular, a great supply of honey is obtained; and where this tree abounds, in connection with a profusion of white clover, there is the apiarian's true *El Dorado.*

Common red clover, that seems so very inviting, is perfectly useless to the honey-bee, as so many thistle heads; for the reason, that the probosis of the bee cannot penetrate the nectaries of this flower, owing to their great length.

As a fall source of honey, nothing can equal *buckwheat;* the honey, however, is not of so fine a flavor, as that made from white clover, let who will assert to the contrary. It is much darker than that gathered in June or July, from other sources, and it will not command so high a price, as that obtained from other flowers. Buckwheat affords a supply of honey for about four weeks, and every bee-keeper who is a farmer, should sow plentifully of this article, for the twofold purpose of the grain, and its advantage to his bees.

Some people imagine that the vicinity of extensive flower gardens, is highly beneficial to bees; such as the gardens of gentlemen residing in the immediate vicinity of large cities, where almost every flowering plant and shrub that adorns both hemispheres may be seen. This is a mistake. Bees do not frequent such places at all, unless it be to visit a few of the common order of flowers. *Roses, pinks, tulips, carnations, dahlias, &c.*, have no attractions for this insect; but where these things exist, may generally be found, a rich harvest for them. As I have already said, the blossoms of *cherry, nectarine, peach, apple,* and *pear trees*, are their first resource; then comes the mantle of white clover, from which a speedy harvest is reapt.

I recommend no especial crop to be sown for bees, as a source of honey, except *buckwheat;* and this is profitable of itself, to say nothing of the honey that it yields.

CHAPTER XVIII.

HONEY DEW.

THIS is a substance that is supposed by some to be an exudation from the leaves of trees; such as the oak, laurel, bramble, poplar, willow, &c. Other naturalists have considered it a substance that falls from the atmosphere.

Mr. Ducarne, a foreign naturalist of distinction thus speaks:—"You know what honey is, which the bees collect with so much ardor from flowers, but you do not, perhaps, know that there are two kinds; one, which is real honey, being a juice of the earth, which proceeding from the plants by transpiration, collects at the bottom of the calyx of the flowers, and thickens afterwards; it is, in other words, a digested and refined sap in the tubes of plants; the other, which is called the honey dew, is an effect of air, or a species of gluey dew, which falls earlier or later, but in general a little before and during the dog days. The dew alights on the flowers, and the leaves of the plants and trees, but the heat operating on it, coagulates and thickens it, whilst, on the other, the honey which falls on the flowers is preserved a much longer time.

"Those persons who have not viewed the honey dew fall, like myself, have asserted, that it is nothing more than the sap or juice of the plants, which, in hot weather, experience, perhaps, a greater fermentation, and by which it is forced through the leaves. In contradiction to this, I assert that it is perceived much better in the morning before the sun has been able to dry and harden it. These persons are, however, deceived. *I have not only seen this honey dew fall a hundred times* in the form of fine rain on the leaves of an ash, but I have also showed it to others, and the globules were most distinctly perceived." Whether this substance be an atmospheric phenomenon, or an exudation, or secretion of certain trees and shrubs, is of little consequence to the apiarian, beyond merely satisfying his curiosity.

I will now give a little testimony on the other side of the case.

"I have long adhered to the opinion," says Mr. Knight, "that the honey dew deposed on the leaves of the trees, was only an exudation, although the form of the globules scarcely bore any resemblance to each other, but were rather an imitation of a species of rain. On examining more minutely different trees, on which the honey dew was apparent, chance led me to the discovery of a holm-oak, on which the honey dew had recently appeared, and in its primitive form, which is that of a transfused humor. The leaves were covered with several thousands of globules, or small, round and compact drops, which, however, seem to be either touching or intermixing, similar to those which are seen on the plants after a thick fog. The position of each globule appeared to indicate, not only the point from which it exuded, but also the number of the pores, or the glands of the leaf in which this mellifluous juice had been pressed. I assured myself that the honey dew possessed the real color of honey, which alone was sufficient to decide its origin, and our surprise need not be great that exudation is not suspected as the cause."

Thus it will be seen that this subject is but another question of the disputed catalogue that pertains to the history and economy of the honey-bee. Some naturalists contend that there are two sorts of honey dew, neither of which falls from the atmosphere, "one a secretion from the surface of the leaf, and the other a deposition from the body of the aphis." Thus speaks Dr. Bevan. The aphis is an insect that abounds on the leaves of certain trees, at certain seasons, and they are said to eject a saccharine fluid from their bodies, in very small limpid drops, of the consistence and flavor of honey.

It is my opinion that no honey dew ever existed, that was not an exudation from the leaves of the tree. It appears to be so inconsistent, that nature should shower down a sweet mist that can only be perceived on the leaves of shrubs or trees. Why do we not perceive the bees gathering it from stones, and other substances, as well as from the leaves of trees, if it were an atmospheric mist or dew?

There is no doubt of such a substance existing, and on this account, the proximity of diversified forests to the apiary is beneficial.

CHAPTER XIX.

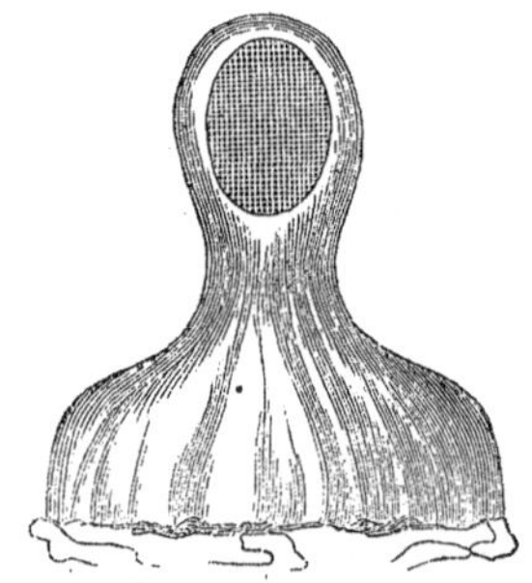

BEE-DRESS, ETC.

The above cut represents a bee-dress or head covering, made by inserting a piece of wire-cloth in a muslin cowl, that reaches down over the shoulders, securing the neck by buttoning the coat over it, as high up as possible.

There are various ways of protecting one's self in operations with bees, but there is nothing superior to that shown in the cut. A veil is often used, or a piece of musquito netting is thrown over the head; but such things are but temporary and imperfect security. If the bee-keeper is accustomed to hiving swarms, and if he be one that is favored by bees generally, and has but an occasional use for protection, he may be as careless of exposing his person to their stings as he chooses; but every apiarian who has frequent use of a protection, and especially if bee stings produce much pain and swelling, should have perfect security against their attacks.

There is a class of persons who are seldom stung by bees, when other people, placed in precisely the same situation, would not escape without paying dearly for their temerity. The reason of bees showing this partiality, is merely owing to the odor of different people's breath. Bees are very quick to take offence when approached by a person whose breath is unpleasant to them.

In consequence of the breath being offensive, it is best to suppress respiration as much as possible, when holding the head immediately over them, or when the breath would be likely to be scented by them.

The principle advantages in a bee-dress on the foregoing plan are, that the vision is not obstructed; a person being able to look through the wire cloth almost as plainly as if no such thing intervened; and the facility that is offered for free respiration. The wire-cloth need not be so large as to cover the whole face, but simply over the eyes, if you please. It is sewed in with ordinary linen thread. Wire-cloth of the kind used for this purpose, may be obtained in cities and large towns, at the agricultural stores, or at bird-cage makers. It should be quite fine and pliable, and the suitable size may at once be known by ascertaining whether you can see distinctly through it when held close to the eyes.

When this dress is put on over the head, the coat should be thrown off sufficiently to allow the lower folds to fall down around the neck and shoulders, when it is to be raised and buttoned up under the chin. Over this head-covering a hat is to be worn, of a little larger dimensions than usual, and one that is kept especially for such occasions.

The lèngth of the dress may be diminished some, from that appearing in the cut; say six inches below the neck does very well; and it may be made without contracting it around the neck as the engraving represents, if you choose to do so. It is the most simple thing to make imaginable. All you have to do is, to get a little black or dark-colored muslin, cut and make it in the form necessary, sew in the wire-cloth, and it is done, costing about one shilling and sixpence.

When the head-dress is made, you then want a stout pair of woollen mittens or gloves, with an old stocking leg, five or six inches long, sewed on to the opening of each; and with these on, and drawn up well under or over your coat sleeves, and with your eyes peeping through the wire-cloth and coat buttoned up to the chin, you will feel like encountering the whole of the bees in your apiary, as if they were so many flies.

It sometimes happens, that when the bees find the head and hands invulnerable, they will descend and crawl up the legs of your pantaloons. It is best to wear boots, and in some cases, when many bees are placed in a situation where they might chance to get into the legs of your nether garments, it will be requisite to tie strings around to prevent such a result.

Woollen gloves or mittens should always be used, as the stings of the bees can be easily withdrawn from such; whereas, they cannot from buckskin or leather, which causes the death of every bee that perforates them.

BEE-STINGS—HOW CURED, ETC.

The venom or poison of the honey-bee, is very active, rather more so than that of the wasp. The fluid is of a transparent nature, and when applied to the tongue imparts a sweet taste. It is not necessary that the fluid should be imparted from the sting of a bee, to produce pain and swelling; the puncture of a needle, with the fluid on its point, would produce precisely the same effect.

The activity of the poison depends somewhat on the temperature of the weather. During the heat of summer it causes much greater inflammation, than in the winter season.

Some persons are much more affected by stings than others; this is owing to a peculiar state of the system or blood, as it exists in different people.

The only *positive* and *immediate* cure for a bee-sting,

that I have ever heard of, that may be depended on in *all* cases, is TOBACCO. This remedy was recommended to me as an infallible cure; yet I had but little faith in it, still I tried it, and as I supposed, properly, and found little or no benefit from its use. I reported its failure to cure in my own case, to my informant, and he stated that I had not applied it thoroughly, as I ought to have done;—that he was certain that it would be an effectual cure, never having known it to fail in a single instance, when correctly applied. The next time I got stung, I applied the tobacco as directed, and found it to cure like a charm! The manner of applying it, is as follows:—Take ordinary fine-cut smoking or chewing tobacco, and lay a pinch of it in the hollow of your hand and moisten it, and work it over until the juice appears quite dark colored; then apply it to the part stung, rubbing in the juice, with the tobacco between your thumb and fingers, as with a sponge. As fast as the tobacco becomes dry, add a little moisture and continue to rub, and press out the juice upon the inflamed spot, during five or ten minutes, and if applied soon after being stung, it will cure in *every* case. Before I tried it, I was frequently laid up with swollen eyes and limbs for days; now it is amusement to get stung.

There are various other reputed remedies, such as ammonia, (spirits of hartshorn,) salœratus diluted in water, cold water alone, and earth mixed with water, and applied to the puncture, and various other alleged cures, all of which I have tried, and found partially ineffectual.

Ammonia is excellent to allay swelling, and if a cloth be saturated with it, and applied to the wound, it will extract the most of the poison, and at the same time take off the skin. I had the entire skin of my forehead taken off by it on a certain occasion, when stung over one of my eyes. I felt a most powerful burning, but being determined to effect a cure, I bore the pain with patience for several hours, when I found that the swelling had abated; but I had lost the skin of my forehead, which was far worse than the sting. It will do very well to apply it in this way, if not left on too long. Occasional bathing with ammonia, is a good way to apply it.

Another remedy is an onion sliced in two, and well covered with fine salt, and bound on the part affected. This is a very good antidote.

Some apiarians insist that nothing is better than *cold water*, quickly and freely applied. This is a remedy that is always at hand.

The tobacco, however, is the great panacea, and I hope that none of my readers will refuse to try it from prejudice. Let every bee-keeper have a small paper of this weed handy, where, in case of being stung, or any of his family, he can apply it without delay. I can cure any sting, no matter how bad, in *five minutes,* with tobacco, so that one would not know that he had been stung, from any sensation of pain that would be felt after its application.

It is always best, as soon as stung, to search for the sting, and extract it, as it is generally left in the flesh by the bee, it being barbed like the shaft of an arrow, and it meets with so much resistance, when the bee attempts to withdraw it, that she is forced to leave it behind.

I would observe in regard to exciting bees to the use of their stings, that the apiarian should never repulse an attack; no matter how many nor how furious they may dash at you, when performing some operation that excites them, always keep calm, and pay not the least regard to their anger. If you find that they are coming rather too hot, discretion may dictate a retreat, even when fully protected with a bee-dress and gloves; and let your retreat be slow and cautious; and if any dense shrubbery be at hand, run your head into it for a minute or two. I have been exposed to their attacks, when their excitement was so great, that they would dash against the wire-cloth with the violence and sound of hail against the windows, in a storm; at which time the odor from their venom-bags was very strong. It is best to retreat, when such a crisis exists, for a few minutes, and let them cool down a little.

When you have a pair of thick woollen mittens or gloves on, you need not fear in the least, that you will be stung. Let as many bees attempt it as may please, do not withdraw your hand suddenly from any position, when covered with them, even if they try their best to sting you; as I have already observed, that any sudden motion tends to arouse their anger.

CHAPTER XX.

SWARMING, ETC.

The primary causes of swarming may be said to be an instinct natural to the bee, which teaches her to extend and propagate her species. This is a wise and universal influence of nature, that pervades all animate creation.

In order to insure this desired result, nature has had recourse to harmonious causes and effects, that produce the ends desired. The only way in which the honey-bee can increase and propagate her species by multiplying families or colonies, is by sending off families as pioneers, to find shelter and protection for themselves; and to insure this, there must be certain causes that operate to force out swarms, even against their wishes.

In order that the reader may arrive at a proper insight of this subject, I will make a few remarks on the general features of breeding, and the particular influences brought to bear on the queen of every family, in the spring of the year, when all measures tending to produce emigration are put in operation.

In the first place the queen commences her great laying

in March or April, according to the state of the weather. If the weather be very mild, she may sometimes commence as early as February; but subsequent cold weather generally intervenes and puts a stop to further laying for a while. She continues to lay eggs in moderate numbers, until about the first of May, when she produces from 100 to 200 eggs per day, for a few weeks. It is at this period that she decides, or, perhaps, her workers decide, whether any emigrant families shall be sent off. They reason thus;—can all the tenants of this hive that now exist, or those to exist hereafter, find room to labor here to advantage? Whether it be the queen that decides this important question, or her subjects, we can never know; but this we do know, that if the space within the hive be such as to afford room for all the family to labor to advantage, it is decided positively and irrevocably, not to send forth any swarms, and no royal cells are constructed! If, on the other hand, the increase of the family will be such as to be unable to find suitable accommodation at home, it is as positively decided, that one or more swarms shall emigrate; and the royal cells are constructed, in which to rear the queens that are to go forth with them, or with all except the first. Thus it will be seen, that the *size* of the hive settles this question entirely.

If it be decided that a family or two can be safely spared, and still leave a populous stock behind, it becomes necessary to create a large number of drones, to ensure the impregnation of the young queens, as has already been fully illustrated. Coeval with the laying of drone-eggs, which generally takes place from the 1st to the 10th of May, is the construction of queen-cells. From five to ten royal cells are usually commenced, and the same kind of eggs that produce ordinary workers, are laid in these royal cells, from which queens are produced, by a different treatment and food, as I before illustrated. The eggs are not all deposited in the royal cells at once, but on different occasions, so as to mature about the time that they will be wanted to go off with swarms. There is generally a superabundance of young queens matured, so as to be on the safe side, and guard against any casualties that may ensue. This is exactly according to human reasoning and human judgment, to provide a few over the exact number of any particular thing desired, where the least risk of loss may appear.

A young queen is never suffered to leave the cell until the *first* swarm has departed with the *old* queen at their head. If any of these young scions of royalty should be ready to emerge from the cells before the swarm is ready to issue, they are kept prisoners therein by the workers until the swarm has departed! Here is one of the most wonderful features of the economy of the bee. Nature has implanted so deadly a hatred of rivalry in the queen bee, that she seizes her own offspring, as soon as a young queen emerges from her cell, and thrusts a deadly sting into her, without the least compunction. Again, nature has ordained that the old queen should go off with the first emigrating family. This is just as it should be. The old queen's impregnation being effectual from the season previous, she is ready at once to go on with the increase of her family; whereas, a young one would suffer the casualties of delay in her impregnation, and thus endanger the existence of the colony; and the issue of more than one swarm being a precarious matter, it is a wise dispensation in the nature of this insect, that the old queen is compelled to leave the hive with the first swarm. I say *compelled*, yes, actually compelled to go forth! Never was there an instance known, where she remained behind, and a young queen took her place. The reason lies here;—the moment a young queen is matured and commences *piping*, that is, says *peep*, *peep*, which may often be heard, it is because she is in duress, or in other words, she is held in confinement, and fed by the workers. When this takes place, the old queen is aroused, and in her anger, she attempts to get at the royal cells to destroy the young queens that are ready to emerge, and she is restrained by the workers. In her desperation and agitation that seems to dementate her, finding that she is not permitted to immolate her young, she rushes out of the hive, calling in her train a portion of the family, being resolved to remain no longer, where her authority is rendered nugatory. It is not wholly the loss of her absolute authority that causes her to depart, but it is also a fear and dread of encountering her rivals to the throne, that also has an influence in causing her to rush from her tenement. When the time arrives for her departure, she commences a sudden vibration of the wings, and rushes over every part of the combs with the utmost speed, and her subjects, in her trail, catch the impulse, and a commotion ensues within, that beggars all description. When the notice is fairly given to the whole family, the queen rushes towards the outlet, and if in her passage she happen to pass near the royal cells, the workers mistaking her intention to leave the hive, for a rush at a young queen, seize and hold her a prisoner. In the meantime the word has been given out to swarm, and away go the workers, as if ten thousand of their deadliest enemies were on the chase. They cluster as usual, but in a few minutes they miss their queen, and all is confusion again. They return to the hive. This is the reason of swarms sometimes issuing without a queen.

If the queen pass near no royal cells, containing young queens ready to emerge, she goes off, and then all is peaceable again.

Now, we will follow the condition of things in the hive after the old queen is departed. The workers at once go to the oldest of the young queens that is kept in durance, if there be more than one, and say "madam, you are at liberty to come out." She comes forth, strong and full of fire and energy, and at once assumes the helm. She, in turn, also scents out her sisters in royalty, and if permitted by the workers, she would fall upon them and slay them while yet in their cells.

We now come to a crisis where all future swarming rests upon the decision of the workers at this juncture. If no more swarms can be spared, the workers immediately give up the guarding of any more royal cells, and the queen that has just assumed the reins of government, being the oldest and strongest, rushes upon all the young queens that may be ready to emerge, as well as upon those in the embryo state, and destroys them also; consequently, no more swarms can possibly take place that season.

If, on the other hand, it be decided that other swarms shall issue, then all other young queens are kept confined as long as possible; and the same causes that drove off the old queen, may also force her successor to depart; but it sometimes happens, that half a dozen young queens will mature about the same time, which are difficult to be kept in confinement, in which case, they are guarded by the workers, and at the proper time, some one of them gives the notice to swarm, and several queens rush out in the general melee. This accounts for more than one queen being sometimes found in a swarm.

A permanent stop to swarming may be occasioned by a few days of rainy weather, occurring just at the time when a family ought to issue. It happens thus;—as the young sovereigns increase in age and strength, the workers find the greatest difficulty to restrain them in their attempts to destroy each other, and they often become wearied out by the delay in issuing, when the weather is long unfavorable, and giving up their royal charge to their own wrath and hatred, it is not long before all are killed save one. Here I would remark, that nature has so wonderfully ordered the attacks that queens make on each other, that in no case are *both* killed in the same combat; if it were not so, many families of bees would be liable to perish.

The season of swarming is a season of peculiar interest to the apiarian. It is at this season, that he looks for a reward of his labors, in the increase of his families of bees. Aside from the profit accruing from an increase of families, there is an interest—I may, perhaps, say a *charm*, attending the issue of a large swarm of bees, to the apiarian who takes a deep interest in the domestic economy of this insect.

When the cry of "*bees swarming*," reaches one's ear, he drops all and runs to the scene. If he be at the well with a bucket of water half way to the top, perhaps, down it goes, with many a hard thump against the stony sides, and away he goes. If he be in the field at the plow, he stops his team when the sound strikes his ear, throws down his whip and is off. When he arrives at the scene, he beholds the heavens darkened with a revolving mass of bees, and thousands still rushing from the hive! Mark the slow and beautifully-undulating circles described, as the bees hover around the apiary, in order to give time for all to join the swarm! Now, a portion of the living cloud quickly and thickly revolve around yon slender branch, where a few bees are already clustered! Now the whole mass, as by magic, draw closely around, and settle thick and fast. The far-extended cloud that but a moment before seemed to cover the area of an acre, now, by some mysterious command, whirl in the space of a few feet, preparatory to clustering in a solid mass! Now all are clustered save a few straggling bees, that seeem to be undecided whether to join the emigrants or return home. There they hang in the form of an inverted cone, with their heads up, enough to fill a peck measure.

"What are they hanging there for?" says a bystander who has never seen a swarm issue before.

"They always cluster in this way, sir, preparatory to taking their flight to the forest, or such places as they would seek for a home; provided, that we should not tender them one."

"Is the *king*-bee among 'em?"

"There is no king-bee, but a *queen*-bee is among them, without doubt. If she were not, they would not remain so quiet as you perceive them to be, but would be seen running to and fro, in wild consternation; and when satisfied of her not being present, they would quickly return to the hive whence they issued."

"But how can they know whether she be with them or not, since she is but one among so many thousands."

"They have the power of communicating this knowledge, which is almost instantaneous. When on the wing, a certain noise produced by the wings, will immediately bring a swarm, extended over many rods of space, to a focus, where the queen may be. When in a cluster, or in the hive, her presence is quickly communicated from one to another. As the General-in-chief gives the word of command to his aids, from whom it rapidly passes down the lines, until the whole army knows the orders of their commander, so is a knowledge of the presence of a queen-bee imparted to the legions under her control."

"Why, sir, the history and economy of the bee must be quite interesting. I always supposed it was a dry subject. Really, I should like to hear a little more about the wonderful instinct of this insect, unless you fear this swarm will depart if not soon hived."

"With pleasure, sir. I'll just put up this canvas screen to keep the rays of the sun from them, since they cannot bear an intense heat in such a situation, and then I'll try and see if I can find the queen, and we will see what effect her removal will have upon them, if I be so fortunate as to find her."

"I'm delighted, sir, at the prospect of seeing her."

"I think I'll find her with the feather end of this quill —don't be alarmed, they don't sting during swarming-time, unless fretted a great deal. *There she is*! I have her now,—I'll throw this handkerchief over her to hide her from the swarm, or they will follow her at once."

"Let me have one look at her, if you please.—How long and slender she is,—*black back, and yellow under her belly, and—*"

"There, sir, now look at the commotion! Do you see how they run up and down the branch in every direction, as if in search of something. Now they begin to leave it, and if I don't return the queen, they will all leave soon.—I've returned her. Now mark the effect."

"They keep up the commotion, and the buzz of their wings yet."

"They will be calmed soon. Now see how they begin to re-form! They are aware of her presence now, sir. A few minutes more, and tranquillity will be restored. Now all is perfectly quiet again."

"I'm astonished! I had some important business in town to-day, but I'll put it off until to-morrow, since I should like to see you hive this swarm. I'm determined to purchase the first swarm of bees I can find, and I should like to see how you perform the operation of hiving."

"John, bring a clean hive, a table, blanket, chair and brush. Nothing can give me greater pleasure, sir, than to entertain my friends in this way."

"You are *very* obliging, sir."

HIVING.

"In the first place, I take a perfectly clean hive, and rub a very little honey around the inside, with a small sponge that I keep for that purpose. I am not positive as it does any good, yet I am sure that it does no harm, and since it is reasonable to suppose, that the bees will like it, I continue in the practice. Various other things are recommended to dress hives with, but I pay no attention to them, since I never lost a swarm on my plan, and I could not have done better if I had pursued other people's plans. The great object is to have clean, sweet hives. Dressing them with the leaves of certain trees, or of herbs, is entirely useless. If the swarm be clustered within six or eight feet from the ground, which is generally the case, where many low trees and shrubs exist, I place my table under them and spread a blanket over it. I then place my hive in such position, that the bees may be made to fall directly before it, and within a few inches of it. I then raise the front of the hive with a block of wood, as you perceive, so as to give the greatest facility for the bees to enter rapidly. Having done this, and being protected by my bee-dress, I take a chair to stand on, and with a brush, such as this, I am prepared for the operation. I will now show you how I do it. With one hand hold of the branch, and the brush in the other, I give a sudden jar to the limb, and down they fall before the hive; and the small portion that adhere to the branch after jarring it, I brush off, and all that take wing again, follow their companions below There, they are all on the table now, except a few hundred, that will soon settle with them in front of the hive."

"That was done very dextrously. Now they are running into the hive, I perceive, except on this side, where a portion are clustering on the outside of it. That is not right, is it, sir?"

"I'll brush them off, and with the feather end of this quill, I'll soon make them disappear."

"Ha! ha! now they scamper—now they go in. Do you leave the hive here, or remove it at once to the stand?"

"It is not a matter of importance, whether it be removed as soon as the bees become quiet, or left here until evening. As a general rule, all swarms that come off in the morning should be removed to the stand as soon as they become quietly hived, and swarms that issue in the afternoon, may be left until evening before removing them. The reason for this course is, that a swarm issuing in the morning, will become so accustomed to the locality when the hive is left unremoved, that more bees are lost the next day, when its situation is changed, than would be, if the removal took place immediately. I do not wish you to understand me, that any bees are actually *lost* in either case, since they return to the parent hive, when they cannot find their new tenement."

"I understand you, sir, perfectly well, and I feel under obligations to you for thus explaining the manner of hiving, &c. Good morning, sir."

So much of the features pertaining to swarming, I have put into the words of others, in the form of a social dialogue.

The foregoing cut, as the reader perceives, represents the actual operation of hiving a swarm of bees, as I am accustomed to perform it, when the swarm does not cluster too high.

A blanket is necessary to spread over the table to ease the fall of the bees upon it. When they cluster so high as to require a fall of five or six feet, I put the blanket on doubled, or throw a bag over the table, and then put the blanket or table cover over it.

Some apiarians first brush or shake the bees off the branch into the hive, which is held under them bottom upwards, and then set it down upon a table with one or more of its sides raised on blocks to admit such bees as are out. This is a good way where the bees do not cluster conveniently for shaking or brushing off before the hive. There is no difficulty in making the bees enter when made to fall on a table before it. They will run towards a hive when several feet from it, on their seeing an opening for them.

HIVER.

When bees cluster upon the branches of trees, too high to admit of being hived in the foregoing way, a temporary *hiver* may be used to advantage. It is made by taking three light, thin boards about ten inches wide, and 18 inches long, and nailing them together in the form of a triangle, with both ends left open, and sundry auger holes bored through the sides, near the centre. An iron strip is then secured to it, with arms extending along two of its sides, and a short shank projecting, which is made fast to a pole. This hiver may be raised by the means of the pole to any usual height that bees cluster, and by the use of additional joints to the handle, secured with ferrules, it may be raised to any reasonable height. It will be necessary to make the hiver as light as possible, in order to handle it conveniently. All that is to be done when bees cluster beyond the reach of ordinary means, is to place the hiver over them, with some of the holes in contact with them, and in a few minutes they will enter it, when it may be taken down, and the bees shaken out on a table in front of a hive intended as a permanent residence, with one side raised about an inch, and they will enter speedily.

CLUSTERING ON THE OUTSIDE OF THE HIVE.

It often happens that a swarm is inclined to cluster on the outside of the hive, rather than enter immediately. This is caused by the heat being insupportable within, or from the queen being outside. In the latter case, nothing will cause the family to remain quietly inside until the queen is made to enter. The remedy is to brush the bees off gently, either with a soft brush or the feather end of a quill, and give them every facility for entering, and also as great a circulation of air under the hive as possible, by raising it on blocks. Every precaution should be taken to keep the hive in the shade.

RINGING OF BELLS, AND OTHER NOISES USELESS.

When a swarm issues, no jingling of bells, or the rattling of tin pans should be indulged in, in the least. This custom originated from the cottagers of Europe, residing in communities making a practice of ringing bells, or thumping on tin pans when a swarm of bees issued, so as to know who the owner was; since swarms issuing from the premises of one cottager would frequently cluster on the grounds of another.

SWARMING PREVENTED BY EXTRA ROOM.

The apiarian may at any time prevent swarming, by affording the bees extra room below them. For instance, take a hive filled with bees, and nearly ready to throw off a swarm, and place it over another hive of the same diameter, with a passage-way through it, and the bees will soon destroy their young sovereigns in the embryo state, and no swarming will take place.

It is sometimes advantageous to thus prevent the issue of swarms, when the owner does not wish any further increase in the number of his families, as the larger the body of bees together, the greater is the quantity of surplus honey produced; yet this argument does not apply to increasing the size of hives, except temporally by nadiring or supering.

Boxes may be constructed of half the usual depth of hives, with both ends open; and in the month of May, before swarming, one of them may be placed under the hive, where swarming is to be prevented. These boxes should be made of the same dimensions of the hive, so as to make a close joint where they come in contact. In the fall, a wire may be drawn through between the connection and sever the combs, and the bees in the lower section will return, and the family will be very populous, and probably highly prosperous; yet but few more bees will exist in February, and March following, than would exist, if a swarm were suffered to issue; but the labors of the extra number of bees existing during the summer, are not lost, as the honey and wax in the nadir will testify.

It is very important to know how far supering, or placing boxes over the hives, will prevent swarming. I have never found supers eight inches deep to prevent swarming with me; neither have I found the boxes in chamber-hives, to prevent it in the least. I always let the bees into the supers in April, and I get my regular swarms. If supers should be placed on of the full size of the hive, swarming would be likely to be prevented; but there is nothing certain to prevent an issue but nadiring, or collateral hiving, in such a manner as to throw two hives into one, or nearly so, from the capacity of the openings between them.

STRONG FAMILIES ALWAYS RECOMMENDED.

I cannot too deeply impress on the bee-keeper's mind, the necessity of keeping very strong families in all cases, if possible. I will illustrate this point, as follows: a family of 15,000 bees is supposed to occupy a hive one foot square in the clear, or inside. That family will, if left undivided, lay up sufficient honey in one season, to carry them safely through the winter, and have *forty* pounds surplus that goes as profit to their owner. Now, what would be the result if they should be divided into *four* families; each with a queen at its head, and placed in hives of the same size of the original one? There would not be a drop of surplus honey stored up by either family, and in all probability, they would not be able to exist through the month of November, unless fed, much less through the winter. It is on the principle that "in unity there is strength." Four rods when together may not be broken, but take each separately, and the whole are easily rent asunder. The philosophy of the failure of four families of 3,750 bees each, to gather as much honey as one family of 15,000, lies here:—it requires nearly as many bees to remain constantly at home, in each of the *four* hives, for the purpose of keeping up that degree of heat that is necessary within, that it does in the hive where the whole 15,000 reside; consequently, it follows, that in one case, perhaps, 10,000 bees would be constantly on the wing, and in the other case of the four separate families, not over 1,250 could be spared from each, making only 5,000 bees as the actual number of gatherers employed by the whole of them.

In consequence of this state of things, more bees are lost, by a desire to increase our families too rapidly, than from any other particular cause. It is truly said, "that experience is the best schoolmaster;" and I have paid pretty dearly for my knowledge. Feeling anxious, one season, to increase the number of my families, to the greatest possible extent, I divided my largest swarms, and some families that did not swarm but once, I drove out, and made two families, where but one existed before. This course well-nigh ruined my whole apiary; and I would briefly say, always be on the safe side in regard to the strength of your swarms, and never grieve when but one issues in a season. I assure you, that one large swarm is enough. Never think of dividing them, unless you are positive that they contain two or more issues. There is not much danger of your dividing swarms or families, unless you have had experience in such matters. I will illustrate this point soon.

DIFFERENT SWARMS APT TO CLUSTER TOGETHER.

When different swarms issue at the same time, they will almost invariably cluster on the same branch. This arises from an instinctive predominant principle in bees to congregate in as large families as possible. It is not necessary that each family should issue at the same instant; since a swarm already clustered, will be followed by a swarm sallying forth half an hour later; and another coming forth before the previous two are hived, will be sure to mix with them. In extensive apiaries, there is much difficulty attending the union of swarms in this way. In such cases, it is best to have everything at hand ready, and hive each swarm as quickly as possible. When the weather becomes fine after a rain, and it is probable that several swarms will issue at once, it is advisable, when a swarm commences issuing, to sprinkle the rest of the hives with water, from a watering-pot. This will keep them back a few minutes, until you can hive the one already clustered. I now speak of very large apiaries, where from 25 to 100 hives exist. Every precaution should be taken, to keep the hives that have just received the swarms, as much out of sight as possible, as it frequently happens, that a swarm will follow another, after being hived, if a portion of the bees cluster outside, where they may be seen.

When several swarms do get together, making, as I have known, a barrel full of bees; and perhaps a dozen different swarms, then the apiarian is in no very enviable predicament. I heard of a gentleman whe had 200 hives or families, and when they came out and clustered together in this way, he hived them in a *barrel*, and in one season the barrel would be filled with combs, and contain several hundred pounds of honey.

TIME THAT SWARMS REMAIN CLUSTERED.

The length of time that swarms will quietly remain upon the bough where they cluster, if not hived, is a matter of importance to every bee-keeper. There is not that necessity for hurrying, as if one's life were at stake, as some people imagine. If the weather be unusually hot and sultry, and the swarm cluster where it is fully exposed to the rays of the sun, and it be between the hours of *eleven* and *two*, you cannot be too quick in securing them; but if they issue in the morning or in the afternoon, when the air is cool, or if they are fully shaded, let the time be when it may, you can hive them at your leisure. I had two swarms issue, some few years ago, when the weather was not oppressively hot, under the following circumstances:—I had occasion to be absent from home at a period when no one was on my premises who could hive bees. One swarm came out about 10 o'clock, and the other about 11 o'clock. They remained quietly clustered until half-past 3, when a most violent thunder-storm arose. The wind blew a gale, and the rain came down in torrents, for the space of an hour. At 5 o'clock, I returned, and found both swarms clustered as at first, and not a bee had been lost by the force of the wind and rain. This case is a fair criterion of what may generally be expected, when swarms are left unhived. They will often remain 24 hours, and sometimes they will adhere to the branch where they cluster, until every bee perishes, or returns to the parent hive. From my own experience, I am led to believe, that the length of time that swarms will remain where they cluster, depends, in a measure, upon the fact, whether a general supervision be extended over them by the owner; that is, whether he is constant in attending to the little duties pertaining to the apiary; such as brushing away the webs of insects, keeping everything in order, feeding a weak swarm here, and attending to the wants of a family there, and by his daily presence, manifesting to the bees, that they are not left to provide wholly for themselves. As " the ox knoweth his owner and the ass his master's crib," so is the little bee sensible of the fact, that a hand is ever ready to provide for her necessities. Though you cannot change one iota of her natural economy, that she has brought down through thousands of generations since the creation of the world; yet if you but extend kindness to her —if you feed her when famishing—if you remove impediments to her prosperity, that she cannot perform, she remembers your attention, and learns to place her trust in you. This is a prominent feature of every being that depends on man for protection. It is an attribute of Him who created all.

The mandate went forth at the creation of the world, "that as man looketh to me, and I extend an outstretched arm over him; so shall every living thing be subjected unto man, knowing that he provideth for them in the day of their necessity." Taking this view of the case, it is not unreasonable to suppose, that if one seldom goes to his apiary, and pays little or no regard to the wants of his bees, they will, in swarming, have no idea of being provided with a tenement; and consequently will, perhaps, take to the forest much sooner than under other circumstances. I have been led to this conclusion, from hearing of many swarms departing to the woods, in cases where I knew that no attention was paid to the wants of the bees generally; and from the fact, that during the many years that I have kept this insect, I never had an instance of a swarm departing, except one that clustered on the sunny side of a tree, where the thermometer was about 140° in the sun, hot enough to roast them; and I should not have lost this one, but I was not present until half an hour after clustering, and they took flight just as I arrived. I say that I have had but *one* instance, I have had two; the other was a case where the person hiving them, used salt too freely in dressing the hive, as I shall narrate hereafter.

THE QUEEN GENERALLY ALIGHTS FIRST.

The above cut represents the commencement of clustering. The queen generally selects the branch to cluster on, and wherever she goes, the family are sure to follow; sometimes, however, the bees cluster while she is on the wing, and she follows the swarm, but such cases do not often occur. If it so happen, that the queen becomes fatigued, and alights on the ground or in some place, where the bees cannot readily observe her, they will cluster without her, and remain a few minutes only, when every bee will return home to the parent family. Queens are often forced to alight before a suitable branch is selected to cluster on, in consequence of the shortness of their wings, not enabling them to fly with the same ease as their subjects.

NECESSITY OF PREPARATION FOR HIVING, ETC.

The above engraving shows how the apiarian should

be prepared to hive his bees without delay, when he can do so; because they cannot be hived too soon, and you may be too late. Always have a common table handy, and a blanket or an old table cover, where you can lay your hand on it, at a moment's notice. A brush as appears in the apiarian's hand, in the cut, should also be at hand. Your hives should be in order, and perfectly clean, and always a few more of them than you may actually require, perhaps, should be constructed. Hives that have been previously used, are as good as any, if perfectly clean. Boiling hot water should be freely used in cleaning old hives, and the joints well drenched to kill the ova of insects.

Bees, when swarming, are quite docile, seldom using their stings, unless in windy weather, when fretted a great deal by the branches or leaves of the trees flapping against them. The person on the right hand of the cut, who is defending himself from their attacks, foolishly commenced parrying and striking at a stray bee, that came around his ears in rather a menacing attitude, and by so doing, he brought a dozen around his head, breathing vengeance for the affront. He will know better next time.

Bees are very particular about the weather when they swarm; and the first swarm more particularly, as the old queen goes off then, and she has more experience than young queens in such things. A calm, sunny day is chosen for migrating generally. If a storm arise at the time swarms are expected, and continue one or two days, or longer, the first fair day will bring them out;

provided they be ready, and the storm has not continued so long as to break up their arrangements, as before illustrated. Some writers assert, that bees never swarm when high winds prevail. This is a mistake. They will wait for pleasant, mild weather, as long as they can, and then let it be windy or not windy, they come forth on some occasions. During the month of June last, (1848,) I had a swarm issue when the bees were almost blown to the ground, before they could cluster. There had been four days of the most windy weather that I ever knew at that season, and on the fifth day, while the wind was still rushing past like a gale, this swarm issued.

The time intervening between the first and second swarm, is from *nine* to *fourteen* days, but generally about the ninth day; between the second and third, *seven* days; and if still another issue, on the *second* or *third* day thereafter.

If a storm arise immediately after hiving a swarm, and continue long, the bees must be fed. A piece of empty honey-comb placed under the hive, and filled daily with liquid honey, or syrup made of sugar will answer the purpose. The bees, in such a case, must be confined, so as to exclude their neighbors, as honey is quickly scented; much sooner than sugar made into a syrup.

SYMPTOMS OF SWARMING.

No positive symptoms showing when a swarm will actually issue, can be given. Huber has had a great deal to say in regard to what he terms "*piping*," as being a symptom showing when a family will sally out; but this indication is not to be depended on, from the fact, that not one bee-keeper in ten, will ever be able to distinguish this sound, amid the *hum* of a populous family in warm weather.

Piping, as it is called, is simply the notes of the young queens, that are held prisoners by the workers, as before described, manifesting their desire to obtain their freedom, and the noise emitted sounds somewhat like *peep, peep;* and when such a sound is heard on a calm evening by applying the ear close to the hive, and in actual contact therewith, it signifies that a swarm will issue soon, if the weather be favorable.

The only general criterion by which we can judge whether a swarm will soon issue, is from the following circumstances, viz:—

If the hive be full of combs, and the bees find difficulty in getting into it at evening, a swarm may be soon expected, any time after the 15th of May. If large clusters of bees hang out at evening, then the symptom of swarming is still stronger. If no swarms have come out on the 1st of June, and the aforesaid symptoms exist, it is almost morally certain that one will depart very soon, unless the weather be cold, damp or windy. When a swarm has issued, clustering out indicates that another family will take their exit, but not much dependence is to be placed upon the apparent populous state of the stock, for any issue except the first. If the weather be very warm, the apiarian will be liable to be greatly deceived in regard to the actual population of his parent hives, and he will think it a pity, perhaps, that more swarms are not sent out, when, if such a result were to take place, it would be the ruin of his apiary, to a great extent.

There are instances, when all signs in regard to swarming, may fail. Every bee-keeper, or at least, many, have watched their hives with a deep interest, during the swarming season, wondering what keeps their bees, as it were, spell-bound to their tenements. Large clusters will hang out, night and day, and the swarming season will pass away, and still there they hang, apparently without doing any labor, save an occasional departure to the fields to supply the wants of nature. These things will often occur, and the reason why no swarms issue is, that a failure has taken place in the production of young queens, or when produced, they have been slain from some cause, as I have already explained. The only alternative with such over-populous families, is to form an *artificial* swarm, or suffer them to remain as they are. When bees cluster in large inverted cones on the under side of the bottom-board, it is well to place a few handfuls of grass directly under them, as they often fall to the ground in the night, or during the prevalence of a storm.

SEASON OF SWARMING.

In the latitude of New York, the usual season of swarming is from the 15th of May, to the 10th of June. in higher latitudes, for instance, that of Boston, it is a few days later, perhaps; and in more southern districts it is somewhat earlier. Occasional swarms may issue in April, and also as late as July, and even in October, instances are found of such a result. When a swarm issues in October, it embraces the whole family; and it may, perhaps, be more properly a *desertion*. The two instances of this nature, that occurred in my own apiary, and before alluded to, came out in the month of October, leaving both honey and larvæ behind. Powerful, indeed, must be the cause that forces a family of bees to leave their domicil at such a period, and depart on the wing to an uncertain destiny. The bee has the same natural attachment for its young that pervades all animate nature. When a piece of brood-comb is extracted containing larvæ, the bees adhere to it with the utmost tenacity; and the cause of such an unfeeling, and apparently uncalled-for desertion, may appear strange to one not having a tangible idea of the true reason. My opinion on this question is, that the hives being but partially filled with combs, not over one-quarter part, and there not being over one-tenth the number of bees that constitute a populous family, the idea of wintering in a place where no warmth could be generated by them, and having had a foretaste of what was to come, in a few cold days previous to their departure, with the entrance all around the hives open as in summer, they foresaw that death must ensue if they thus remained, and having, probably, sent out an embassy to find some hollow tree in which they would be less exposed to the rigors of the weather, they departed. The reader may recollect, that I stated, in a previous allusion to this singular desertion, that I succeeded in hiving both swarms; and in a few days they rushed out again, and I was unable to stop them. Here is a cut representing a swarm taking a flight to the forest, high above the trees.

When bees are determined to seek a home for themselves, they revolve in a mass, gradually getting higher and higher, until the coast is clear, and then their flight is rapid; yet sometimes they may be followed for half a mile. The best remedy for bringing them down is, to throw fine sand or water among them. When one of the swarms issued, before spoken of, I seized a pail of water and a dipper, and I made the water fly among them like a real shower. Before I had used the first pail of water, they had got some twenty rods from the apiary. In the mean time I sent for more water, and I at last succeeded in bringing them to cluster on the branch of a cherry tree, about twenty feet from the ground. Being called away for a short time, I suffered them to remain, and when I returned they were gone, and to my satisfaction; since my only motive in stopping them was, to experiment on the application of water in such cases.

In regard to the actual danger of the aforesaid two swarms perishing, had they remained in the original hive, I would observe, that they were the smallest swarms that I ever had; consequently, I am not able to say positively, whether they would have survived through the winter or not. I have wintered swarms that did not contain over a quart of bees in December, with perfect security, and I am inclined to believe, that had I lowered down the hives, and allowed but a single place of entrance, and had fed them freely, they would have lived through the winter. The two swarms alluded to, would not probably have made more than about a pint in bulk, if left until December or January. There was about a quart in each when they departed.

SWARMS CONSIST OF BEES OF ALL AGES.

The question has often been asked me, if swarms are not composed entirely of young bees? My answer is, that they contain bees of every age, from the old bee of the season previous, coming into existence through the summer and fall months, down to the young bee that never before ventured a rod from the hive. There appears to be no discrimination on the point of age. A promiscuous sally takes place, and the majority are young bees, as a matter of course, since four fold the number of old ones existing, are produced every spring.

SWARMS ISSUING HAVE NO HABITATION SELECTED.

Much has been said in regard to bees selecting a habitation before issuing from the hive. It is supposed by many persons, that previous to a swarm issuing, an embassy is sent to the forest to select some hollow tree or other tenement in which to reside. That such cases sometimes do take place, I have no doubt, but that a selection is made in every case, is not a fact. I think that the attention paid to the wants of the bees by their owner, as before alluded to, has something to do with this matter; since it appears that they who pay no regard to their bees, at any season, lose swarms very frequently, while they who are constantly paying the little attentions to them that good management demands, seldom lose any.

I have often observed single bees, during the season of swarming, entering the knot holes of my stable, singing a merry song, and carefully examining them for no other purpose, I presume, than to find a domicil for some swarm soon to issue. It appears natural in the bee to send out scouts in this manner, and where a forest is very near, filled with hollow, decayed trees, the securing of swarms is attended with much more trouble than in other situations. It often happens, that a hole in some old building is secured by them for a tenement. A lady of my acquaintance, who is very fond of attending to bees, informs me that on a certain occasion, a swarm issued from her apiary, and without clustering, proceeded slowly across the fields to the house of a neighbor, about half a mile off. Their progress was so slow, that she was able to follow them all the distance, without losing sight of them. They entered the house at an aperture under the roof. She returned home, and in a few minutes another swarm issued, and took the same course without clustering; and she again followed after them, and just as she arrived, the last of the latter truant swarm were entering the same aperture where the first swarm entered. This is rather a singular case, since it is very rare to have swarms issue and depart without clustering. Clustering seems to be necessary, in order to congregate the whole family prior to the journey, when a flight is contemplated. Some apiarians have recommended decoy hives; that is, empty hives placed about the apiary, in which it is supposed swarms may enter. I have tried this experiment, but have never found it to succeed.

BEES COMMUNICATE ON THE WING.

While bees are swarming, they have a peculiar power of imparting information from one to another, while on the wing. It is this power that calls, at a moment's notice, the bees that cover many rods area to a focus, when it is decided to cluster. I had a singular circumstance occur, the last season, of this nature. I had several swarms issue and cluster on the same branch, at the same time. I divided them in three parts, and hived them separately, thinking that if I should happen to get a queen in each, it would save all further trouble. I placed the hives in different situations, and in the course of a few hours I found the whole together again. I then took a small swarm that issued the day previous, and placed it where the hive stood that contained the three swarms, which was filled inside and covered outside with a perfect sheet of bees, and as quickly as possible, I shook out about half of its contents alongside of that containing the small swarm, and then I ran with the hive thus emptied of half its contents, and set it in the place where I took the other from. My object was to force a division in some manner, if possible. I remained a few minutes watching the result. Presently the bees that had been shaken down at the side of the hive, commenced rising on the wing, and mixing with a small part of the other portion of the joint families, that hovered in the air, and soon a trail was formed from one hive to the other, and in fifteen minutes, every bee was at the hive in the new situation, some rods off; the knowledge of its position having been communicated from one to another, on the wing. It is only in cases of large numbers of bees being on the wing at the same time, that a communication can be effected by them.

PROPER MODE OF SEPARATING SWARMS.

The foregoing experiment was out of the regular order of my usual course, in effecting a division of swarms; yet such a separation, or rather an equalization of families may be made. Two swarms, the one very large and the other very small, may be managed thus;—at evening, when the bees are all in, and quiet, take the large family from its position, and supply its place by the small one, at the same time, causing a portion of the large one to fall alongside of the hive containing the small swarm. The bees will readily enter, and join the family within. The hive containing the large swarm should be placed in the position that the other family occupied. To effect such results, it may sometimes be necessary to perform the operation of mixing the bees on a blanket, spread on the ground. One hive may be be set down, with one side raised half an inch, and the bees from the other falling near it, will enter at once. The time to effect this equalization should be soon after sunset, and neither swarm should have issued beyond a day or two previous.

During the next two days, still larger numbers from the large family, will gather to the small one; since they go out to the fields from the new situation, and return to the old one. The object of forcing out a portion at once, is to mix with the other family, and partially destroy the peculiar scent by which bees from one swarm recognize those of another. It is not advisable to force out many, as a very large portion entering so as to outnumber those already there, might cause trouble with the queen, as a strange queen coming suddenly into the midst of a great number of bees, not of her own family, is at once seized by them, and held so close a prisoner, that suffocation is liable to ensue. I do not recommend this way of separation of swarms, unless it be in cases where one can well spare a large portion of its numbers, and the other cannot possibly thrive without an accession to its strength.

The true and proper way is by adopting the principles of *artificial* swarming, as I shall soon explain. It simply consists in attaching a piece of brood-comb containing larvæ, in two or more hives, according to the number of families that may be desired to make, and then dividing the bees, placing them in the hives with the brood-combs, and if no queens happen to be among them, they can make them, as I have before stated. If the division take place on the day of swarming, brood-combs should be placed in each hive; but if left until three or four days after being hived, they should be placed in the empty hive only, since the full one would probably, already contain combs and eggs; and there would be no necessity of disturbing the queen pertaining to it. The manner of division in this latter case is thus:—at any time of day, a small portion of bees should be gently shook out of the hive containing a double family, in such a position, that they will enter the empty hive, which is to be placed in the full one's position; then carry the full one to a new situation, not less than ten feet off, and during the two succeeding days, the empty hive will gain strength as before described, until a respectable family accumulates.

UNION OF SWARMS.

A union of two or more small swarms may be effected on the day of swarming, without any trouble. All that is to be done is, to make one mass of them, and the extra queens will soon be slain, and the bees will work as one family. A union may be effected in this manner, at any time within a week after hiving the bees, but there is this difference attending it; one of the hives is to be removed, of course, and during the next two or three days, a large portion of the swarm that had become accustomed to the place where the removed or emptied hive stood, will return to that place, as they sally forth to the fields from their new domicil, and will be lost.

BEES LIABLE TO CLUSTER ON THE APIARIAN.

The above cut represents a servant of mine, with a swarm of bees clustering on his neck and back. Perhaps the reader may think that this comical scene is but visionary; yet I would assure you, dear reader, that such things do actually occur: and every bee-keeper is liable to be placed in the same unenviable predicament.

The queen, as I before stated, is sometimes quite heavy and unable to fly, and it may happen that in case of not meeting a proper shrub or branch to alight on, she will perch on the first thing that may come in her way; and if she alight on any part of the person in attendance, the whole swarm will follow so speedily, that there is no help—no evading it. All that one has to do is, to stand still, and bear it like a philosopher, not attempt to run away, as poor Sambo did. There is not the least danger in such a case, if one will only be quiet. It will call forth all the presence of mind that he is possessed of, without doubt. In such a case, another person should bring a hive and hold it over the bees, resting it in some manner that will give facility for them to enter, and in a few minutes, they will all take to the hive.

SAMBO'S FIRST TRIAL AT HIVING.

When I commenced keeping bees, I was as green in the business as the most ignorant. I gave directions to my servant to dress the hive with salt and water, having heard that this was good. When I returned at evening, (I resided on Long Island, and came to New York daily,) he informed me that a large swarm had issued, and he hived it, and in a few minutes they rushed out, and that was the last he saw of them. I turned up the hive, and behold! he had rubbed salt enough on the sides of it, to pickle a pig. There was no wonder why the bees departed. The next day, when I returned, he came up grinning, "I've got 'em, sir, this time, and a mighty large swarm it is, too," said he. What, another swarm out, inquired I? "Yes, sir, here it is, and to make *sure* of 'em, I cut off the branch, and put it into the hive, and then tied a cloth over it tight, leaving one little air hole for 'em to breathe, sir." Sure enough, there was the hive tied up like a band-box on a journey. I flew around and made the necessary preparations to remove them, and called my wife out to witness the first swarm, as I should remove the covering, and turn up the hive to our admiring eyes. I got the whole family around me, as assistants in the operation, all as eager to witness the bees as myself. I gently raised the cloth. The end of the branch protruded under the hive. I raised the hive on one side very slowly, with my heart in a flutter of excitement and anxiety; I expected soon to hear exclamations of delight on all sides. Higher and higher the hive was raised, expecting every instant to catch a view of the swarm clustered at the top. Presently it gained a point where the eye reached the summit, and such a sight! Reader, what do you think it was? *Not a single bee was there!* If ever the mercury went down suddenly in my blood, it was then. I was nigh breaking the hive over Sambo's head, but the old fellow was useful, and I let him off with a good scolding for having duped me. The bees had escaped one by one, through his air-hole, and had returned to the parent family.

GRAPE-VINES SUITABLE TO CLUSTER ON—ARTIFICIAL CLUSTERING BUSHES, ETC.

A grape-vine seems to be a particular favorite for bees to cluster on. I had several large vines near my apiary, on Long Island, and I have frequently had every swarm during a season, cluster on them. When the leaves are large enough to afford shade, they are inclined to cluster on them more, than when the vines are partially bare. They often cluster in peach, apricot, cherry and apple trees; and not unfrequently on currant bushes. Where no small trees exist of the size of ordinary peach trees, some kind of small tree should be set out, and I know of none better than the peach. I have tried artificial clustering shrubs or bushes with the most perfect success. In the spring of 1848, I removed my apiary to a place where not a tree or shrub existed of a suitable size. I took a dozen of the poles used for sustaining dahlias, about six feet long, and to the end of each I fastened a green cedar bush about one foot in diameter, and eighteen inches long; being the tops of small cedar trees and shrubs found in any quantity, in the woods. I drove down these poles in different places around the apiary, some two rods apart; making the bushes stand from four to six feet high. When the swarms issued, they selected one of these bushes. I had twenty-six issues, and every one clustered in the same way; and seemed to like them better than trees, as they afford the best security against the bees falling. They generally clustered around the centre of the bush, and when they became dry and faded, it made no difference in regard to the bees clustering on them.

APPEARANCES AT THE MOMENT OF ISSUING.

Perhaps some of my readers would like to know how a family of bees is affected at the first movements in swarming, since very few bee-keepers witness the first outward excitement among them. I stood looking at a hive last June, and presently a mass of bees came surging over and over, out of the hive on every side, like the froth of a pot slowly boiling over. The bees ran up the sides a few inches, making a complete mass all around it, and then for a moment, the most phrenzied excitement took place, running in every direction, with a speed unnatural to them. Then they commenced rising on the wing, and in one minute, the air was darkened by the mass that issued.

TIME OF DAY TO EXPECT SWARMS TO SALLY FORTH.

From 7 o'clock in the morning to 6 o'clock in the afternoon, they may issue, but generally from 10 to 3. In very warm weather swarms frequently issue as early as 7. This is about all the specific guide any one can have.

CHAPTER XXI.

ARTIFICIAL SWARMS.

The art of forming artificial swarms has been known for many years. Shirach is reputed as the original discoverer of it. It is simply based on the power that the workers possess, to convert any worker-egg or larvæ under four days old, into a queen, as I discussed in Chapter II. This being an established fact, it follows, that if a queen belonging to any family be removed, and leave behind eggs or larvæ, of a suitable age, the bees will rear another queen, and proceed in their labors, as if nothing had occurred to them.

I will now illustrate precisely what would take place within a hive, if a queen were suddenly removed. We will suppose that the whole interior of the hive is fully exposed to our view. We now remove the queen suddenly, and without molesting the workers in the least. All is perfectly quiet, and perhaps may remain so several hours, since no alarm has been sounded. The workers take it for granted, that the queen is still there; and although they do not see or feel her, yet it is presumed that she is somewhere about the hive. We will arouse them, and on the approach of danger, their first impulse is to be assured of the safety of their sovereign. We take a rod and rap smartly on the hive.—The bees now begin to run speedily over the combs—the excitement increases, and they are now fully aware of the queen's absence. Hark! what a tumult and roar within! How eagerly they traverse the combs, as if in search of something. Six hours have now past, and the excitement is dying away. Here in this cluster of bees, the rudiments of a queen-cell are already laid, and within twenty-four hours one of the larva of the unsealed cells will be removed to it, and in about 12 days a queen will issue.

If a strange queen should now be offered to them, they would not receive her kindly; but would cluster around her in such numbers as to suffocate her, in all probability. But if we wait 24 hours, and then offer a new sovereign, she would be welcome, and would be treated with the respect due to royalty, since it requires 24 hours to cause a family of workers to forget their queen.

On an occasion of endeavoring to unite two small artificial swarms, where I was apprehensive that one of them was without a queen, I attempted driving them out by the aid of smoke, and the family that I supposed to be without a queen, did possess one, and she rushed out of the hive and alighted on the under side of the brim of my hat; I seized her, and placed her under a tumbler until the next day, when I turned up a hive that contained another artificial swarm, which, to all appearances, had no queen; and having laid the hive on its side, I placed her majesty close up to the bees as they clustered on the combs, and stood awhile to watch the result. At first they did not seem to notice her, but presently, two or three workers extended their antennæ towards her, and at once appeared excited; and in a few minutes, a dozen or more gathered around, holding her a close prisoner. She endeavored to extricate herself from them, and very plainly articulated the sound of *peep, peep.* I heard it as distinctly as I could hear a chicken's call. She soon disappeared in the mass of bees, and I saw nothing more of her.

This was a queen that had been reared from the worker larvæ, that I had introduced into her hive, some three months before.

The benefit to be derived from artificial swarming, is in cases where families send off no swarms, as often occurs, from causes already narrated.

The method of performing the operation is as follows: take a clean empty hive, and attach at the top, in one corner, a small piece of brood-comb, containing both eggs and larvæ; at least larvæ in cells *not sealed ove1* The younger they are the better, and even eggs alone are sufficient, since an egg of to-day, will become larva to-morrow or next day. The manner of attaching the comb is as described at page 201, and the utmost care is requisite, in order to cement it firmly. When that is done, on a fine day about 11 or 12 o'clock, as the greatest number of bees are out at those hours, you then remove the stock with surplus numbers, to a new situation, as far off as convenient, and not less than ten feet; and if you can brush off a portion of bees, that cluster on its sides before it be removed, you should not fail to do so, at the same time, having the empty hive in the full one's place to receive them. By such means, a nucleus is at once formed around the brood-comb. If there be no clustering outside, you should manage in some way, to get about a quart of bees to enter the empty hive at once, as your success depends upon it. If no other way offer, you must turn the full hive bottom upwards, and set the empty one over it, and with a rod strike the lower hive for a few minutes, when a portion of its inmates will have entered the empty one, and clustered on the comb above. The two following days will add large numbers to the artificial swarm, besides all the bees that are out in the fields when the operation is performed. If the full hive be removed without leaving bees enough behind to form a nucleus around the brood-comb, the bees returning from the fields finding an empty hive, will run around in distraction, and perhaps depart entirely; but if they see a cluster already in the hive, however small it may be, they will join it; and after the first six hours, they will go to work and rear a new queen, and in the fall there will be no difference between this swarm and one that has issued in the natural way. Artificial swarms must be large, or there is a liability of their not rearing a queen, for the want of sufficient animal heat within the hive to develop her; and also on account of the delay attending their own natural increase, from having to wait two weeks, at least, before a queen will be ready to commence laying.

TIME OF YEAR TO MAKE ARTIFICIAL SWARMS.

Artificial swarms may be made as soon as the 15th of May, and as late as the 15th of June, with safety. Later than this period, would be attended with some risk of success; not only an account of the lateness of the season for gathering honey; but also for the reason, that there would be no certainty of a sufficient number of drones existing when the queens mature, to effect her impregnation as speedily as the case demands.

ARTIFICIAL SWARMS MADE WHOLLY BY DRIVING OUT.

In certain cases it may be best to drive bees enough out at once, to form the swarm desired; and if the queen of the old family be driven out, there is no necessity for attaching a piece of brood-comb in the empty hive. The best method of performing the operation in this way is, to drive out about two-thirds of the bees of the stock, and the queen is almost sure to go out with them, before so large a portion of the family departs. When this is effected, place the empty hive now containing the queen and two-thirds of the family, in a new situation; and the old hive where it has always rested. The bees in the new hive, with the queen, will be content, of course, and they will commence comb-building in a few hours. A small portion of them will return to the old stand, enough to equalize the families, which is just what is desired. The bees in the old hive will miss their queen, and make a great uproar about it for a few hours, but will finally go to work and rear a new queen from the larvæ that was left when the old one vacated, and the larvæ left will mature at the proper season, if the weather be warm, and thus increase the family, and at the end of three months, both families will have filled their respective hives with honey, wax and bees.

DIRECTIONS FOR DRIVING AND DISLODGING BEES.

Notwithstanding that I have already stated, how bees should be driven from one hive to another, in brief, yet I am aware, that I cannot give too plain, and explicit directions for this operation, which to the inexperienced bee-keeper, must at first be attempted, with any feelings but those of pleasure. In the first place, you must be perfectly protected by a bee-dress and gloves. No half-way work in such matters. Provided the weather be warm and favorable, it may be done at any time of day, if you are to drive out a full swarm; and if much clustering exists outside, the time when the least exists is best. You take the full hive and turn it over carefully, setting it down on the ground or table, on its upper end. The empty hive is now to be set over the full one, making a close joint, so that no bees can escape; and I would here observe, that all hives should be of the same diameter in every apiary, in order to effect such operations with ease and facility. Having effected the junction, an empty hive is to be placed where the full one stood, as a decoy, to keep the bees that return from the fields from entering the neighboring hives, until the operation is performed. A cloth is now to be tied around the joint, where the two hives meet, to make it as dark as possible within the hives. This done, the lower hive should be rapped smartly with a small rod on all sides, for the space of ten or fifteen minutes; when, in all probability, half or two-thirds of the family, with the queen, will have ascended into the upper hive, and clustered there in a compact and quiet body.

HOW TO CUT OUT BROOD-COMBS.

This is a job that is not coveted by the amateur apiarian; yet it must be done, where artificial swarms are to be made; and when once performed, it is quite easy to do. All that is necessary is perfect protection, that does not obscure the vision,—a steady hand, with courage and perseverance, and all obstacles dwindle into insignificance.

In the first place, I will introduce to your acquaintance, a couple of very handy instruments, that every bee-keeper should possess.

One is a long knife, with an edge on each side, and sharpened at the end, so as to admit severing combs from their attachments with facility. The other is a long steel rod, with a two-edged knife at the angle, for the purpose of cutting combs horizontally. One edge of the blade is turned directly towards the reader, and the other from him. The length of the rod and handle, should be about 18 inches, and the length of the blade at the angle, an inch and a half. The diameter of the blade on its flat side, should not be over a quarter of an inch, as it is often to be inserted between combs, where the space is not over three-eighths of an inch.

These two instruments are useful in cutting out pieces of brood-comb, as well as for various other purposes, that every apiarian will see the necessity of, many times in a season. If you possess nothing of the kind, you must take the sharpest and longest knife in your kitchen. You have a carving-knife, of course, and if it be a little curved at the point, the better. Take this and put on your bee-dress, and I will then tell you what to do. All ready, I perceive. Now, take this stone, and sharpen the point of your knife on it. Sharpen it on a whetstone? No: if you're going to learn how to cut out brood-combs from me, you must do as I say. I know it makes it as rough as a saw, but don't get into a passion, that is just what I want, it will cut honey-comb better in that rough condition, than it would if it were as sharp as a razor. Now, sir, turn this hive over on its top. Afraid to do it? There, now it is over—is any one hurt? Now take your knife and run it obliquely through one of these centre combs, cutting with the point of the knife only. Can't see, there are so many bees? well, feel your way, then, but cut slowly, so as not to irritate or kill them. Now loosen the attachment at the side, and with your left hand, hold the comb from falling. Yes, take hold of bees and all, they can't sting through your glove. There, sir, what do you think now? The operation is over, and you are alive.

If an ordinary knife be used, a large portion of larvæ

are destroyed by cutting the comb obliquely, more than would be by cutting vertically, and then horizontally, with the knife with an angle, as shown in the cut.

ARTIFICIAL SWARMS FORMED BY DIVISION.

This method of making swarms at pleasure, consists in having hives made in two parts, to divide in the centre, somewhat on the plan of the collateral hive at page 186, with this difference, that where both parts are united, they form a square hive, of the usual size. The two parts should be so constructed, that they may be separated at any time with ease. From necessity, they must be like the lower section of the box-hive, as represented at page 153, that is, when joined.

After being connected, supers may surmount them as on the box-hive, with a couple of holes through each half. The two sides coming together should have the greatest possible open space between them, and not admit of a union of the combs in each half. A narrow, and very thin strip of board should be placed across, at the top, and sunk into the hive its whole thickness; and in the middle, and at the bottom, place the same strips, about two inches wide. This is to be done on one side only. The other half requires nothing. The strips should be very thin, not over a quarter of an inch thick, at most. It will be necessary to insert guide-combs, or the bees might build transversly, or narrow combs across each part, and in that case they would be apt to unite them through the interstices, or passage-way between the two parts. In order to unite them firmly, a hook and staple should be placed on each side, and one at the top, perhaps; and if the hive rest on pins, a piece of sheet iron should be secured to the bottom of one of the parts, on both sides where the union takes place, and made to lap over a half an inch, so that the bottom of the other half may catch and rest on them.

We will suppose, that we have a hive on this principle, full of bees, and we wish to make an artificial swarm. We take another hive of the same kind, and divide it. We then unhitch the full hive, and slowly, and carefully remove one of its sides a few feet, and supply its place with an empty one, corresponding in every particular. We next unite the other empty half to the half of the full one, that was removed, and await the result. The queen will be in one half, but in which, must be proved as follows:—take a rod and beat each hive smartly, arousing the bees as much as possible; and that part which contains the queen will be quite tranquil after a few minutes; but a tremendous confusion will exist in the other, and the bees will run around, under, and over the hive in great consternation. The part where the bees are quiet should be removed to a new place, and the other should be placed in its position. In the fall, two prime families will exist, perhaps, equally as good as if no division had taken place, and no swarm had been thrown off.

ARTIFICIAL SWARMS MAY BE TRANSPOSED.

Sometimes it may happen, that one or two very large, and also a few very small artificial swarms may exist in the same apiary. Such swarms may, within the first two weeks of their existence, and before the queens mature, be transposed with good results; that is, take a very large swarm, and place it in the position of a small one, and *vice versâ*. An equalization of families can be effected in this way, as the bees mix with perfect peace at such times. No other swarms, or stocks can be treated in this manner, without ruinous results to both families.

CHAPTER XXII.

CHANGING FAMILIES FROM OLD TO NEW HIVES.

It is a well-known fact, that bees thrive best, during the first four or five years of their existence in the same tenement. It has been often asserted, that the lack of animation, and of the decay of families, is in consequence of every generation of bees, bred in the same tenement being smaller than their predecessors; on account of the yearly contraction or diminishing of the cells. It is said, that the silken shrouds, that enclose the larvæ are left behind, pressed to the sides of the cells, when the young bees come forth, thus causing them to become gradually smaller and smaller, until there is not room enough for the larvæ to become fully developed, in their natural size and vigor. This hypothesis I must dissent to; notwithstanding it is heresy to do so. Who, among my readers, has compared the size of bees, in different hives, and found some a dwarf race, while the tenants of the other hives were of the full, natural size? Perhaps some among you have *imagined* that you could discover a difference, and so have I, but on close inspection, I found that I was mistaken. I have a hive, in which the bees have resided during ten years, and not a particle of difference in the size of its tenants from those of other hives, can be perceived. An acquaintance of mine assured me, some few years ago, that he had a family which had inhabited the same hive, from generation to generation, *twenty-nine* years, with no difference in the size of its occupants from those of other domicils.

It is my opinion, that the cause of deterioration, is not as above stated; but in consequence of the blackened and vitiated state of the combs, rendering the atmosphere within impure, and having more or less lodgments of the moth to eradicate, from year to year, until the effluvia of the combs operates to the injury of breeding, and through that cause, to the final destruction, in some cases, of the family. Be that as it may, we know that on the fourth or fifth year, it is best to effect a change. How that is to be done is the next question. The "subtended" plan will not answer, for reasons already given; but if we choose to take the lives of our bees in the old way of using *brimstone*, we can destroy our old families every fall, and leave our young ones ; but this method is a cruel and a barbarous one; and wholly unnecessary, to say nothing of the *loss* that the owner sustains by such a course. The method pursued in dividing families by a division of the hive, affords new combs for *one-half* of the tenement; and this mode may be pursued with tolerable success, and two families are made where, perhaps, but one would exist, in the case of *driving* out the bees into new hives. Driving them out, makes the operation perfect, and if not done until a swarm issues, two families are just as certain to result from it as on the previous plan.

The operation of driving out should be performed as follows: The bees should not be disturbed before the fore part of June, in order to see whether any swarms are to issue, and to give an opportunity for as many larvæ to develope as possible. Whether a swarm be sent off or not, it is not advisable to wait beyond the 20th of June, as the bees must have time to lay in sufficient honey for their winter use. At evening, or early in the morning, take an empty hive and surmount the full one, as before directed, winding a cloth around the junction. Then, as before stated, beat the sides of the hive, until all the bees have ascended into the empty one, which will generally be effected, in 15 or 20 minutes; if the old hive be full of bees. If it be but partially filled, they cannot be forced out at all, without the use of smoke.

Here is a smoke-pan that may be advantageously used at times. Tobacco smoke is most effectual in forcing the bees to depart speedily, but anything that will produce smoke may be used. A little cut, smoking tobacco ignited in the pan, with the cover let down, would answer the purpose. If the chamber-hive be used, the boxes or drawers may be removed, and the pan set in the chamber. The smoke will ascend through the holes, and by this means, together with rapping the hive, the bees will be made to ascend; or the pan may be placed under the lower, or open end of the hive, and force the bees up into the box in the chamber, which can be withdrawn, and the bees emptied down at the side of the hive, that is intended for their use, and they will readily enter. In such a case, but one box should be inserted in the chamber, open at the bottom, as I have directed that they should be made. If the box-hive be used, the super will receive the bees, and with more facility than boxes in the chambers of other hives. When the most of the bees are driven out with the queen, the combs of the old hive may be cut out, and the few bees remaining, will join the rest of the family, where new combs will be constructed at once, the same as if the bees had swarmed out in the natural way.

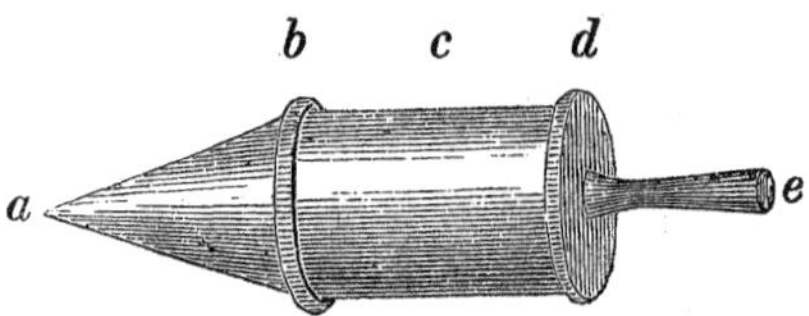

The above cut represents a fumigator, as described by Dr. Bevan. I do not approve of it, but I give it a place, and they who choose can make use of it. *a*, is the funnel, with a hole in the end to let out the smoke; *b*, is a plate extending across the fumigator, perforated full of holes to admit the passage of the smoke only; *c*, is a cylindrical portion of the box, three inches in diameter, and three and a half long, in which the tobacco is placed, *d*, is the lid, which is received into the box when the tobacco has been lighted; *e*, is the tube which should be adapted to the size of the bellows-pipe. The whole is made of tin, having the joinings soldered. It is used by inserting the bellows-pipe in the tube, and then the action of the bellows drives the smoke out through the funnel.

They who are accustomed to smoking, often perform any operation requiring the aid of smoke, simply by directing a few whiffs from a cigar or pipe, into the hive, where the removal is to be effected.

The following cut shows a much better, and less expensive fumigator.

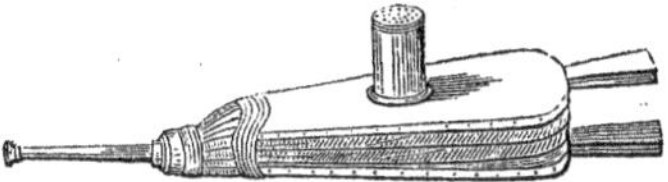

It consists of an ordinary bellows, with a tin tube, about three inches long, and two in diameter, fitted over the air hole. The cover to this tube is perforated with holes; and the air hole is covered with tin, also perforated in like manner. When this apparatus is used, open the tube, put in the tobacco ignited, close it, and the action of the bellows carries the smoke out of the bellows-pipe. This is the most simple and practical fumigator in use.

UNITING STOCKS.

It often happens that the apiarian finds it necessary to unite two of his stocks, or old families. The cause that leads to such a necessity, is frequently from over-swarming, or sending out more colonies than can be safely spared; thereby weakening the parent family so much, as not to be able to recover during the season. When two weak families exist of this character, if they are united, one prosperous family will be the result; whereas, if left separate, both would be destroyed.

The difficulty attending the union of old families, lies in their unwillingness to mix peaceably. There is a certain peculiar scent pertaining to the bees of every family, and especially to old ones. In order to obviate this difficulty, and cause both families to mix without strife, the following plan may be adopted. Take one of the hives at evening, and turn it up, then spread over it a gauze or millinet covering, or anything that will allow a free circulation of air through it; then take the other hive, containing the family that it is desired to connect with the first, and place it thereon, giving the bees of neither family an opportunity to escape; but allowing them sufficient air for respiration. Leave them in this position 48 hours, and at the end of that time, the scent of the two families will have become so blended and interchanged, that they may be united with perfect safety. The process must be by smoke applied to the lower hive, with the use of the rod, after withdrawing the cloth that divides them.

CHAPTER XXIII.

THE SEASONS.

FALL MANAGEMENT.

The months of October and November are the season when the state of the apiary will require particular attention. The hives should be examined, and those not containing honey enough for its occupants to sustain them during the winter, must be fed. An ordinary swarm or family of bees, will consume from 15 to 20 pounds of honey, from October to May. If the winter be very mild, more than this quantity will be required; but not in an ordinary season. The apiarian should be able from practice, to know at once on raising his hives, whether the above quantity exists in them or not. Hives that have been occupied several years, will be as heavy without any honey, as others that have been used but one season, with from five to ten pounds; therefore, an allowance must be made for the weight of old combs, and bee-bread.

FEEDING BEES.

When it is ascertained what families are short of honey, measures should at once be taken to supply them, since the cost of feeding a family, is not one-tenth of its value. The month of October, should be selected for this purpose. If but one or two families, out of ten or twelve, require feeding, it is best to feed those alone; but if there be a general scarcity or lightness of the hives, I recommend feeding the whole in the apiary, at the same time. I am aware that feeding bees is generally looked upon as one of the greatest difficulties attending their management; and rather than attempt it, many bee-keepers suffer their bees to perish. The difficulty is just as great as it is to carry a pail of feed to the pig-pen, and no greater. Do not understand me, by this comparison, that bees will take honey from a trough, as a pig will take meal and water. It only requires a little difference in tendering it to them, however.

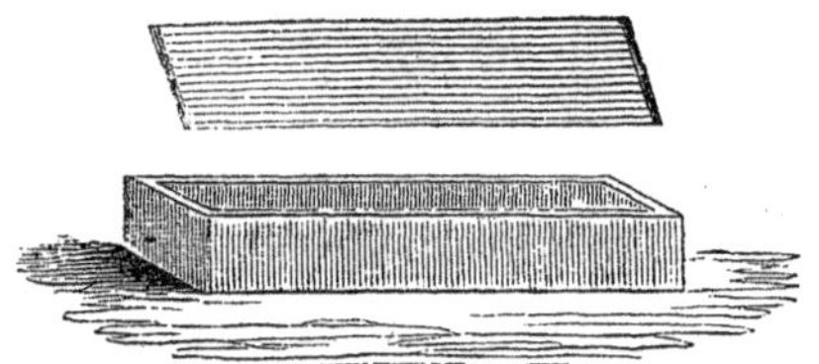

Here is a feeder, with a cover to float on the surface of the honey, when put into it. The box may be as large as one chooses to make it. For an apiary of from two to twelve hives, it should be about 18 inches long, six inches in diameter, and four in depth. The float is made by slitting it with a fine saw, as many times as possible, to within an inch of the end; and the board is half an inch thick. The other end is secured by a clamp or bracket nailed across it. Open the interstices with a knife, by trimming off the edges of the channels made by the saw; then put a couple of little knobs, or nails at each end, in the centre, to serve to raise and lower it; and the feeder is completed.

If honey be fed, it should be such as comes from the W. I. Islands, which is sold at about 62 1-2 cents per gallon, of 12 pounds; or by the cask at 50 cents. If it be thick, and candied, it should be heated to the boiling point, with a little water added to thin it. The only thing except honey, that can properly be fed to bees, is syrup made of sugar. This answers every purpose of honey, and may be made to cost from four to five cents per pound, only. Sugar that sells for five cents per pound, mixed with sufficient water to make the syrup of the consistence of honey, will reduce the cost of a pound of it to about four and a half cents; and from five to ten pounds, will generally be sufficient for the most destitute family; which, at most, will not cost over 45 cents. Thus it may be seen, that for this trifling sum, and perhaps often for half the amount, a family of bees may be kept from famine. The method of preparing syrup from sugar, is to heat it over a fire, until it begins to boil, when it should be taken off, and let it stand half an hour for the scum to settle, and harden, in order to skim it off with facility; and when cool, turn it into the feeder, and put on the float, and set it before your bees. At first, they will hardly notice it, but a few drops should be placed about the feeder, to call their attention to it, and when they once get a fair scent of it, a gallon will be taken away in a few hours, and stored in the cells. This way of feeding, is when the whole colony are fed, rather than select here and there a family. Mild weather in October should be selected for this purpose, and all the food should be furnished that they may require during the winter; and a family will carry enough honey, or syrup into the hive in a single day, if not disturbed by other families, to suffice for several months. If it be desired to feed a single family, it must be done secretly, that other families may not be attracted; and for this purpose a small feeder is necessary. Take a small tin pan, six inches long, four wide, and one or two deep. Make a very light, wooden float, and perforate it full of holes, with a pointed iron heated red-hot. Fill the feeder with honey, or syrup, and place it in the chamber of the hive, scattering a little around, to attract the bees to it; at the same time rapping on the hive to cause the bees to ascend; or place it below, and shut the bees in until it is emptied. Families may be fed at any time during the winter when the sun shines, by having a temporary, or permanent glass door to the chamber, or super of the hive, that will admit the sun's rays pretty freely upon the division board, upon which the small feeder filled with honey should be placed. I once fed a very small family in this way, that had not a drop of honey in the fall. It was no trouble. Every day that the sun shone, the bees were up in the super in great numbers, even in the coldest weather; and in the following spring they increased rapidly in numbers, and soon filled the hive.

Some people are in the habit of merely mixing a little water with the sugar, and not heating it at all, and in this condition feed it to the bees. This is downright ruin to them. In a few days the water evaporates, and the sugar hardens in the cells, so that it is of no more use to the bees than so much flint stone, to say nothing of its destroying every cell that it hardens in for any further use. I recommend honey to be fed in the fall, if convenient, and syrup in the spring.

WINTER MANAGEMENT.

This is a critical season for bees, and their proper management is but very imperfectly understood at this period. The great principle that should actuate the apiarian, is to keep his bees as cool as possible, as I have before inculcated. The practice of burying hives in the ground, and immuring them in cellars, is all wrong. Now and then, a man will have a family pass the winter in this way, without actual ruin; and it is forthwith heralded as a grand discovery—the very best way to keep bees over winter, &c. It is not so. I have neither time nor room to say much on this method of wintering bees, but any place that is not *perfectly dry*, is no place for bees in any season. In cellars, the combs will mould to a greater, or less extent, thereby laying the foundation for the ruin of every family thus circumstanced.

The passages from the lower sections of the hives, to the chambers, or supers, should be left unclosed. This allows the steam or vapor, arising in hives in winter to pass off, and in cold climates, it prevents the accumulation of frost within them, that would otherwise occur.

As I before stated, the hives should be let down in the fall, and the bees made to pass in and out, through the small passages.

Every strong family should have a current of air passing under them, to prevent the bees desiring to come out; and for this purpose, remove the slides to both the front and rear openings. Small, weak swarms, will be kept sufficiently cool, by opening the front entrance only.

When the ground is covered with snow, be particular to confine your bees, if they come out much, by closing the entrances with the zink slides; and as soon as an opportunity occurs to let the bees take an airing, they should have the priviledge of doing so. The hives may be occasionally raised, and the bottom-boards cleaned off, which will aid in keeping the hives free from impurities.

SPRING MANAGEMENT.

This is the season to close the rear opening, or passage-way, perfectly tight, and stop the current of air that passes under the bees during the winter months. All the heat that can be produced from the rays of the sun, will be beneficial during April and May, at least, and if March be a mild, pleasant month, then also. If it be a raw, cold, snowy month, let the rear entrance be open, when it is not necessary to confine the bees, as little or no breeding will take place, if the weather be very chilly and cold, and the cooler the bees are kept the better.

It is supposed by many persons, that the honey-bee passes the winter in a state of hybernation or torpidity. This is a great mistake. I have often seen my bees in populous families, quite lively on turning up the hives, when the thermometer stood at zero. There is a natural animal heat existing all winter, in strong families, even in the coldest weather.

When warm weather approaches, in May, the hives may be raised to afford an opening all round. These directions are given, on the supposition, that the hives rest on pins, and in the fall these pins are lowered into holes in the floor-board made to receive them. Other kinds of hives should be managed as nearly on the same principle as possible.

If the hives are light, the bees should be fed freely in

the spring. A little ale or wine, and a little fine salt mixed with the honey or syrup, is good. A shilling spent in feeding, often produces a dollar before the season is over.

SUMMER MANAGEMENT.

After the swarming season is over, nothing can be done, of consequence, but to keep the apiary free of weeds, and protect the bees, as much as possible, from the inroads of insects. From the 1st of July, to the 1st of September, is the season of spiders and the moth. Spiders will nightly weave their webs around the hives; and the apiarian should almost daily pass around the apiary with a brush in hand, to destroy them. The only enemies to bees we have to fear in this country, are spiders, wasps, king-birds, and the bee-moth. Wasps are of little account. Spiders make sad havoc, if left undisturbed. King-birds will destroy thousands of bees in a season, if no means are taken to destroy them; but all the above enemies united, sink into insignificance, when compared with that terrible destroyer, the wax or bee-moth.

THE BEE-MOTH—HOW ERADICATED.

This insect is of a whitish, or brown grey color, and somewhat smaller, generally, than the ordinary millers that flit around a candle at evening. They are the most nimble insect known. They will dart among the bees, in and around the hive; and before a bee has time to turn her antennæ towards them, they are out of reach. If one attempts to kill a moth when resting by day, on the outside of the hive, by quickly putting his finger on her, the act must be instantaneous, or she is far away before his hand touches the place where she rests. The best way to destroy them, when they can be found outside, is to put on an old mitten or glove, and striking very suddenly with the flat of the hand, will generally prove effectual. They may often be found on the outside of hives during the day, as the only time that they enter is in the evening, or during the night. They generally seek some place where they can pass the day under some board, or any projection that affords shelter under it.

At evening, as soon as twilight appears, they commence flying around the hives, and seeking out such as are not very populous, for their scenes of depredation. Having gained an entrance, they run up the sides of the hive, and at the upper end, or as near as may be, they at first make an incision in the propolis that is used to cement the corners and joints of hives, with their ovipositer, and in the orifice made, the egg is deposited, and so on until they have finished. The heat of the hive keeps the propolis in a soft, pliable state, and it is exactly suited to their wants. In a few days the eggs are hatched, and small white worms emerge. These worms grow very rapidly, and immediately search around for food; and the combs adjoining are very acceptable, filled, as they are, with honey, larvæ and pollen. The bees have an instinctive hatred to these worms, which prevents them from destroying them as soon as hatched.

They do not seem to be aware of the danger that will arise from them, until they commence the destruction of the combs. Having gained a position in the combs, the worms commence weaving a silken shroud around themselves, to protect their bodies, leaving the head only exposed, which is armed with a helmet impenetrable to the sting of a bee. Protected in this manner, they move from cell to cell, eating as they move, having only to thrust out their heads to find food in any direction. Their course is longitudinally through the centres of the combs, seldom appearing on the surface. Their shroud for protection, is carried along with them. Thus it will be seen, how very difficult it is for bees to dislodge this enemy, when a footing is obtained by them.

There is but one way that they can be destroyed, when fully fortified among the combs; and that is, by cementing them in with propolis. This the bees will sometimes do, confining them to very close quarters, and when all the food is consumed within their reach, they perish. On other occasions, whole segments of combs that have become infected, are destroyed by the bees, in order to remove the evil. When the moth gets the upper hand, and the worms begin to increase rapidly, the bees stop all further labors, and the condition of the family is readily known by their inactivity, and from the numerous particles of pollen, comb, &c., upon the bottom-boards of the hives, caused by the progress of this insect. These particles are of a dark color, and are most easily discovered in the morning, before the winds arise, and before the bees commence sallying out. For the purpose of detecting the ravages of this enemy, hives having an open entrance on all sides, either by suspending the floor-board, or resting the hives thereon, with pins at the corners, are decidedly far preferable to those on any other plan. The moth-worm, when having free and uninterrupted sway in a hive, rich in wax and honey, grows to a large size, sometimes being an inch long, and as large as a pipe-stem. A quart of such worms will often occupy a single hive, before all the bees will depart.

INDICATIONS OF THE MOTH—BEST COURSE TO PURSUE.

Every apiarian should closely watch his hives during the months of July and August, and any that show signs of the existence of the moth therein, should be attended to without delay, as the whole apiary might become infested by this pest, arising from a lodgment in a single hive. Every worm, after a few days, must wind up in a cocoon; from which, a winged moth-miller issues, able to produce a thousand eggs, each egg to produce a worm, which, in turn, produces a miller, and so on until a million of worms may exist in one season, from a solitary insect! If the family be weak, and the hive full of combs, where the moth exists, the quicker the combs are cut out, and the bees dispersed the better; or the bees may be driven into a super, by the aid of smoke, and then placed in a clean hive and fed, if the honey season be past, and they will probably survive the winter, if there be a moderate family, and the next season they will replenish the hive in numbers, and be as valuable as any in the apiary. Another way, is to join the infected family to a weak one, that is not yet subjected to the ravages of the moth; the operation to be performed as directed before, for the union of stocks. Do not, on any account, suffer any of your families to become fully destroyed, before you take measures to remove the evil. Who among you, would suffer an animal to sicken and die of a distemper that you know to be liable to spread to the whole herd or flock, and take no measures to eradicate the threatened evil? It would be deemed *insanity* on the part of him who should let such a case pass unheeded; yet the condition of your apiary, when the moth gets the upper hand of a family of bees, is a fair parallel.

POPULOUS FAMILIES NOT LIABLE TO BE UNDERMINED.

There is, however, this difference in the case, every very strong and populous stock or swarm of bees is not liable to be destroyed, being able, by mere force of numbers, to prevent a lodgment being made; and here lies the grand secret of success in the culture of bees; *to ever keep our hives full and populous.* This is the Alpha and the Omega of bee-keeping—the *sine quâ non,* without which, all other measures fail. It is the apiarian's chart—his polar star—the needle that never points but to *success*—the cornerstone, upon which the whole fabric rests.

Reader, have you ever been importuned to purchase hives that were represented to be "*proof against the moth?*" Well, sir, when a perpetual motion, the philosopher's stone, and a north-west passage to the Pacific are discovered, you may believe such a thing possible—not before; and even then, will *I* be an unbeliever.

"How shall we keep our hives full and populous?" says one. I answer, by attending to the correct size of hives to begin with,—not to allow over swarming—to unite weak swarms and stocks, and follow the general rules laid down in this Manual, and you will find success easy.

Swarms are not liable to be attacked by the moth, for the reason, that they extend their area of combs no further than they have numbers to defend them; hence the proof, that populous families can protect themselves. Sweetened water or vinegar, or milk alone, put in white vessels, and placed near the hives at evening, will decoy the moth-miller, and be the means of destroying many. Thus ends the duties of summer management.

CHAPTER XXIV.

PILLAGE OF BEES.

A GENTLEMAN having a field of corn adjacent to his premises, into which his fowls daily resorted, threatening serious ravages; and to stop such a catastrophe, he placed a measure of corn before them, and kept it constantly replenished; the consequence was, that his fowls had no occasion to visit the field. Now, the bees in an apiary, that commence robbing from a neighboring hive, do so from necessity, not from an innate principle of disregard of right and justice; and let the apiarian but place a trough of syrup or of honey before them, for a few days, and all pillaging will come to an end. Some bee-keepers think it very unreasonable, that they should be required to feed their bees, but expect great profits from them, without any trouble or expense whatever. The poor bee is not at fault when she finds her combs empty, and herself in a starving condition. She labors all that she can; but she cannot ward off the storm and the cold north winds that often confine her, when she would gladly be in the fields. But *six weeks* only, out of fifty-two, does this insect have to replenish her hive; the rest of the summer affording but enough for a daily supply; therefore, he who would let his bees perish for the want of food, when a cold and inclement season has deprived them of support, ought to be put on a short allowance himself.

When bees commence robbing their neighbors, the hive attacked should be closed up immediately, on the first evening after the discovery, and remain so a few days. When it is opened, the entrance should be so diminished, that but a single bee can enter at once, and left in this manner for a while. This course will generally prove effectual. Changing the entrance from front to rear, will sometimes cause the marauders to decamp, and it may be necessary, in some instances, to remove the hives robbed, to a new and distant situation; but this should be the last resort. When a hive is being robbed, it may be known by the numerous bees that fly around it, uttering an entirely different sound from that of bees while gathering honey. They seem to act as if they were guilty of a misdemeanor, and show a cowardice in every motion. As evening approaches, they may be seen to leave the hive rapidly, even after twilight sets in. This is the time to close the entrance. Robbers generally come from one family, and they may be discovered by sprinkling flour on them as they emerge, and then watching where they enter.

CHAPTER XXV.

GUIDE PLATE, ETC.

INSTEAD of inserting guide-combs or bars, as Dr. Bevan recommends, to cause the bees to build their combs at proper distances; I recommend the use of an invention of mine, termed a *guide-plate.* It is made of tin, and is one foot square, and sheets may be purchased of just that size. This plate just fills my hives, that measure a foot in diameter. Having ascertained the natural distances of combs, I have interstices cut in this plate to correspond therewith; and previous to my bees swarming, I melt some bees-wax, lay the plate over the roof of the hive on the inside, take a brush and lay on a coat of wax, precisely as the merchant marks his bales and boxes through a plate for the purpose. The bees being hived, follow these traces of wax, in building their combs. I do not suppose that every bee-keeper will obtain such a plate, yet it is a great advantage and benefit; and it will repay its expence ten-fold.

DISTANCES AND THICKNESS OF COMBS.

On measuring the combs in a hive that were regularly made, I found the following result, viz: *five* worker-combs occupied a space of five and a half inches, the space between each being three-eighths of an inch, and allowing for the same width on each outer side, equals six and a quarter inches, as the proper diameter of a box in which *five* worker-combs could be built. According to this calculation, a hive twelve inches in diameter would allow of nine worker-combs being made, and have a little room to spare, since eleven and three-eighths inches is all the space that would be occupied. The diameter of worker-combs averaged four-fifths of an inch; and that of drone-combs, one and one-eighth of an inch.

The tin plate should be cut for worker-combs only; the openings four-fifths of an inch wide, and the space between them, three-eighths of an inch, leaving the two outside interstices five-eighths wide, in order to fill up the space of twelve inches with nine combs. The extra space at the sides, will allow the bees to build two or three drone-combs, which is sufficient.

CHAPTER XXVI.

MISCELLANEOUS.

In consequence of having extended this work much beyond its originally contemplated limits, I find myself compelled to place the following subjects under one head, instead of discussing each in separate chapters, as I would wish to do, had I the space to spare.

VENTILATION OF HIVES.

The only ventilation that should, in any case, be afforded to bees, should come from the bottom of the hive · and in warm weather, too much air cannot be admitted. Here lies one of the principal advantages of raising the hives to allow egress and ingress, on every side of them. It keeps the bees healthy, and in health they are active, and in activity they prosper, and their owner is benefitted by their labors. The passages to the chambers or supers being open, also have a tendency to benefit the bees in the winter season, as before stated; yet I do not consider these as legitimate sources of ventilation. No outside tubes or air-holes should ever be made in a hive above the bottom.

PURCHASE OF BEES.

The months of March and April 's the best season to

purchase bees; yet it can be done at almost any time. If the purchase be made in the fall, all that it is necessary to know in regard to the family is, whether it be populous, and whether the hive contain honey enough to carry the bees safely through the winter. Turning up the hive, will show if it contain a strong family; as it should be full of combs, and the bees should crowd the interstices down near to the bottom. A sudden rap given to it, with the ear quite near, is an index to their strength. A strong family make a long continued *buzz,* while that of a weak one is quick, sharp, and soon over. The weight of the hive will generally show whether there is honey enough within for the winter supply. It should weigh at least 20 pounds over that of an empty one.

TRANSPORTATION OF BEES.

The transportation of bees in the fall, winter or spring, is not attended with difficulty. The bottom-boards should be secured firmly, with sufficient ventilation; and then hives may be placed in a spring-wagon, and transported almost to any distance. They should be turned bottom upwards, if the shape of the hive will admit it. Ordinary box-hives should have the floor-boards nailed on, and then pry them off just enough to admit the air, and the bees will go safely. I refer to common hives, used by those who pay little or no regard to improvement in such things; and which contain no means of ventilation, when the floor-boards are nailed close. In the summer season, it is more difficult to transport bees, in consequence of the softness and weakness of the combs, rendering them liable to break down. Bees should never be removed at this season. In many parts of Europe, the cottagers make a practice of transporting bees from place to place, as the shepherd does his flock, from pasture to pasture, to obtain a fresh supply of food. The bees, in such cases, are in straw hives, which are more easily transported than wooden ones. If hives are to be removed to any distance within a mile, the removal should take place before the 1st of May, if possible. If the distance be very short, they should not, under any circumstances, be left longer than the early part of April, as when their habits become once formed in any particular situation, many will return to the same place, when removed to a new situation within a mile.

COMBS LIABLE TO MELT DOWN.

During very warm weather, if bees are fully exposed to the force of the rays of the sun, there is some danger of combs melting. I never had any melt in the lower sections of my hives, but I have in the supers. I generally protect every hive; but in this case, I left one exposed, when the sun was most intensely powerful, and the loaded combs fell from their attachments in consequence of the heat.

DISABLED BEES.

I ought to have mentioned in the chapter on swarming, that during the height of the breeding season, hundreds of bees may be seen running to and fro upon the ground, endeavoring to rise on the wing, but cannot. To the experienced bee-keeper, this is no news; but I make mention of it for the benefit of those who are inexperienced in bee-culture, and who might, perhaps, be led to think, that a deadly strife was going on in the apiary. Such bees as are seen under these circumstances, are *imperfect* or *disabled,* and come into existence with a broken wing or leg, or possess some imperfection, that consigns them to immediate ejection from the hive.

DISEASES OF BEES.

Long epistles have been written upon this subject, and more, as I have thought, to fill up, and swell the pages of works on the bee, than to benefit the public, by stating interesting facts. I shall simply say, that we need not trouble ourselves in the least, about "*dysentery,*" "*vertigo,*" "*tumefaction of the antennæ,*" "*faux convain,*" &c. All we have to do is, to afford our bees a plenteous infusion of pure air, at the bottom of the hives during summer and winter, and see that *famine* is not at their door, and the foregoing diseases will all vanish from our apiaries.

ARCHITECTURE OF BEES.

The skill and mathematical knowledge exhibited by the bee in her architecture has astonished philosophers and scientific men of every age. It has been fully demonstrated, that the same space occupied by their cells cannot possibly be filled with any shaped vessels, that will either be of greater capacity, or take less material in the construction. There are but three ways in which cells can be built, and have the sides of all equal, viz: *square, triangular* and *hexagonal;* a *fourth* way is utterly impossible. The hexagonal form is superior to either of the other modes, in strength, capacity, and a saving of materials in building; and this form, the bee has chosen! The *bee,* did I say? No: there is a greater Architect than the bee, who has had the guidance and the direction in this matter.

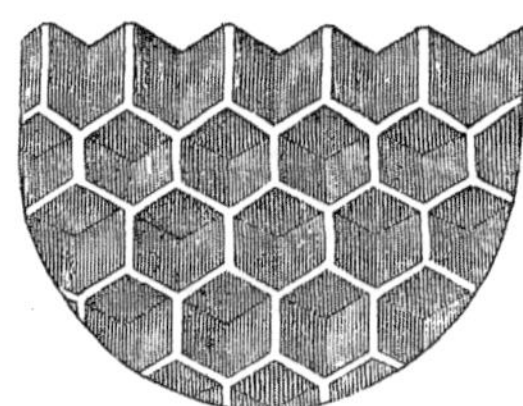

The above cut represents a few rows of cells as they appear when constructed. These cells are not built *horizontally,* but on an *angle.* Here, again, is the most astonishing wisdom displayed. A celebrated philosopher and mathematician being asked of what form a series of vessels united, should be constructed, in order to be of the greatest possible capacity, and take the least possible material to construct them; after a full investigation, he answered; *the shape, hexagonal, and on an inclination of some* 28°, (I think,) *with the plane of the horizon!* The cells of the honey-bee incline from 15 to 28°; that is, the mouths of cells are so much higher than the bases. This inclination is not wholly for the purpose of saving material, but also for the purpose of retaining the honey better. Ordinary brood-combs incline the least, and store combs the most.

One of the most wonderful features pertaining to the construction of combs is, the manner of their junction with opposite cells. Instead of the dividing line between them being a straight line, thus:

---

it is of the following form, and a pyramidical cavity

/\/\/\/\/\/\

is formed at the base of each cell, composed of three triangular rhombs, or portions of wax, at the apex of which, the union of three opposite cells meet.

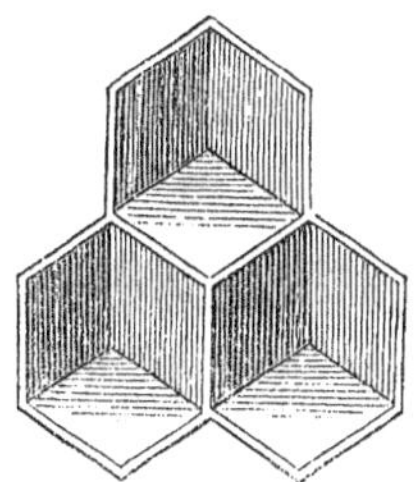

The above cut shows the pyramidal bases of four cells; the apex of one in a cavity, pointing *from* the reader, being the centre, and the other three *towards* him. If the cut were reversed, and the other side made to appear, it would show *three* cavities, similar to the centre one, and that now in the centre, would appear like each of the other portions of the illustration. In the apex of the cavity, the egg is deposited; being admirably adapted to receive it.

I have often closely examined cells to ascertain if I could discover any variation in this rule, and I have ever found the union of opposite cells, to be formed in the centre of the base of every cell thus examined!

It is said, that the hexagonal shape of cells, is not in consequence of any pre-determined action of the bees, so to form them; but the result of the mechanical laws governing the natural pressure of bodies of united spherical tubes, in a pliable and soft condition, before becoming hardened by an exposure to the atmosphere, and that the original form and shape of cells is cylindrical. I must put in my veto to this assumption. I have shaken bees out of hives while in the very act of comb-building; and have had a fair opportunity to examine combs, while yet warm from the internal heat generated by the bees, and I have always found them of the regular hexagonal shape.

The first built, are brood-combs, in order to give an immediate opportunity to increase the family. Small families begin at the side, and strong ones in the centre of the hive. Sometimes a strong swarm will commence on both sides at the same time; and it is not unfrequent, that while a portion of the bees are building from front to rear, another portion will be constructing combs transversely, on the opposite side of the hive, and do not discover their mistake till they meet the other combs, when a right angle is at once formed. This accounts for so much irregularity in comb-building, and it is a strong reason for the use of the guide-plate before spoken of, or of inserting guide-combs on both sides, or in the middle of the hive for very strong swarms, as breeding is greatly retarded by the malformation of combs.

The outer edges of the mouths of cells are strengthened and fortified by a border of wax, much thicker than the sides, which prevents the entrance from being a regular hexagon. This border seems to be of a different material from the substance that the cells are composed of, and of the nature of a peculiar kind of varnish.

The depth of ordinary worker-cells is seven-sixteenth, and that of drone-cells, nine-sixteenths of an inch; and the depth of store-cells, from half an inch, up to three inches. There are but two diameters for the cells of the honey-bee, throughout the whole world! One is for brood-combs, and the other, for drone-combs; the store-cells always being of the diameter of drone-cells. This law is as immutable as the adamantine hills. Take whatever countries you please, England, Russia, China, Africa, Patagonia, Mexico, or the United States, and not one iota difference can be found, if ten thousand families were examined!

The cut on the next page represents a segment of worker-comb, containing eggs and larvæ; also a full-sized queen-cell, and one but partly constructed. The nature of queen-cells having been defined at page 28, it will not be necessary here to say much in regard to their construction. This cut gives a better idea of the natural appearance of royal cells, as they actually appear, than the previous illustration.

The centre of the combs shows a row of cells, in which the egg first appears; then the larvæ just bursting

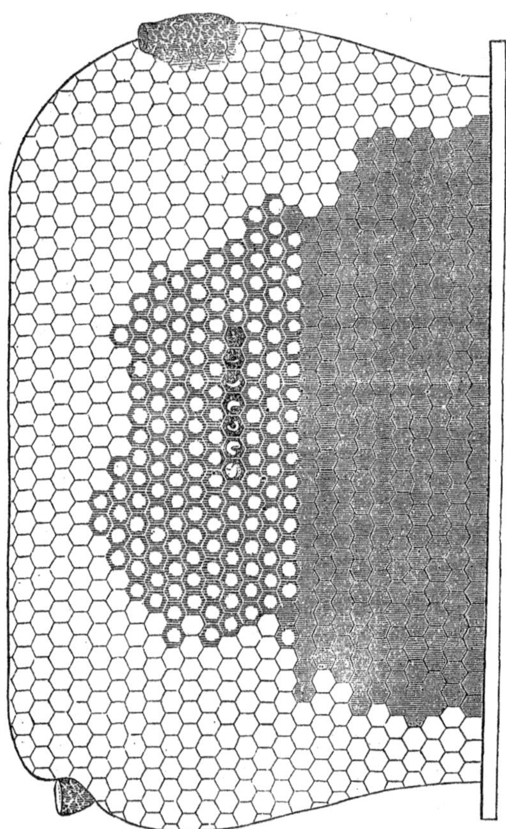

its shroud from the egg; then, as it appears one day older, and so on, until the cells are sealed over, being from the fourth to the sixth day after the deposit of the egg. Adjoining this tier of cells, may be seen those in progress of being sealed over; which operation is effected by commencing at the outer side of each, and attaching numerous small rings of wax, one within the other, until the whole area is covered. Above this section of the comb, containing cells being sealed over, may be seen a portion of cells fully sealed, and from which the young brood emerge, in the course of about fifteen days after being thus imprisoned. On the outer skirts, may be seen the vacant cells, not yet appropriated to any use.

There is no distinction made in a leaf of brood-comb, in regard to what cells shall be used for honey, pollen, or brood. The queen deposits her eggs wherever she finds vacant cells, provided the family be populous; but if not populous, then she takes a very different course, and confines her laying exclusively to the centre of the hive, and to the centres of combs, near the top of the tenement.

In speaking of *store-combs*, I refer to combs built expressly for that purpose, of a thick, irregular form. The whole interior of the hive is used for storing honey, when the cells are not filled with bee-brood or larvæ.

INSTINCT OF BEES.

The knowledge that the bee possesses, as displayed in her achitecture, and general economy, is not acquired by habit, or taught her by those older than herself. She comes into the world, as perfect as she goes out of it. Many are the astonishing instances of foresight and knowledge, of adapting means to ends, that have come under my personal observation; but I can give but two or three of the most important cases on this occasion, which will suffice to show the general features of her sagacity or instinctive powers.

On a certain occasion, I attached a large sheet of comb in a hive, for the use of a family, that I was about driving into it. Some two or three days after the bees had been placed therein, I discovered that a lateral brace had been constructed, from the side of the hive, to the lower end of the comb. This brace was built, in consequence of my getting the comb out of its perpendicular position several times, while turning over the hive to examine the bees. The bees reasoned thus: "He is turning our hive over every day, and our comb bends, and leans over; by and by, it will break off, so we'll build a brace across to hold it!" On another occasion, I laid a sheet of comb, filled with honey, on the floor of the chamber of the hive, covering several of the holes of communication with the family below. I placed it there for the purpose of feeding the bees. A few days thereafter, I was surprised to find this sheet raised three-eighths of an inch, and supported on four pillars built of wax! This was done to give the bees an opportunity to pass up through the holes with facility. The honey had been taken away. But the most astonishing performance that was ever placed on record, as I believe, occurred as follows: Having an entrance to one of my hives, about two inches long, and half an inch wide, that was covered with a thin strip of wood, with a nail at one end, to hold it in its position, I was accustomed to turn up the door or cover, perpendicularly, as I passed the hive and found it closed. The bees had no particular use of this passage-way, as they had abundant egress below; yet, it being warm weather, I kept the cover up, as much as possible. It got so loose by turning it up, that it would often fall down of its own gravity; and not thinking the matter of sufficient importance to secure it at once, I turned it up daily, for about a week, and every morning I would find it down again. At last I turned it up, and out rushed about a hundred bees, and commenced clustering around it in a very singular manner, and I left them and went to town. Not returning until evening, I could not see what the result was before the next morning, when I went out to the hive, and found the cover to the opening so *deeply imbedded in propolis, that it could not be easily removed!!* It appeared that the bees wished to have this hole open, and finding that it was down one day and up the next one, they thought that they would put a stop to it at once, and they did so. I leave the reader to his own reflections on these instances of sagacity manifested in this insect. I could recount many more astonishing operations of the bee, but I am admonished to be brief.

LONGEVITY OF BEES.

The age of workers is generally under *one year*. This fact is easily proved, by placing a family in a large hive, that does not admit of swarms issuing. It will contain no more bees during the succeeding years, than during the first season, or but a few more, at most. Numbers equalling the increase of each season die off before another season approaches. The drones live five or six months, generally, when left to die a natural death; and on some occasions longer, but not often. The queen lives the longest of any of the family, often surviving to sally out at the head of several swarms. Her exact natural age has never yet been demonstrated.

ANGER OF BEES.

The honey-bee will seldom use her sting against any one when not molested, and children, in particular, are exempt. When a bee is aroused to anger, she gives immediate notice of it, and no person was ever stung, unless in the midst of hundreds, excited to vengeance, without having timely warning given him. Every bee-keeper is familiar with the shrill sound emitted, when the bee approaches in a threatening attitude. It is quite unlike the soft song of contentment, that is sung as the bees return from the fields laden with honey. I have never heard of any fatal consequences arising from the stings of bees, except in animals. If a horse or a cow, or any other animal upset a hive, it is generally certain death. In case of being dangerously stung in many places, tobacco, as before stated, is worth more than all other remedies in the world. The duration of the anger of bees, is from three days to a week; and any operation disturbing them much, will not be entirely forgotten, short of that time. Private injuries are seldom resented by them; that is, when molested in the fields.

LANGUAGE OF BEES.

That bees have the means of imparting information

from one to another, is beyond doubt. By what means it is done, has never been fully established; yet it is pretty generally admitted, that it is by means of the antennæ. The antennæ are also the organs of smell, and of recognition of bees of the same, or different families. Besides the antennæ as a medium of communication, a certain noise produced by the wings, is another mode of imparting knowledge, as I alluded to, in regard to families finding their hive, when dislodged, and their tenement, with a portion of the family, being removed to a distant situation. Having a swarm that lay out upon a sheet one night, and exposed to a drenching shower, I found them in the morning with only the outside bees drenched, and the majority were in a condition to be hived. There were several clusters of them, and having made the larger portions enter the hive, I aroused the small ones, within a few feet of it, and as quick as the hive was perceived by them, and a portion of the bees entering, they commenced fluttering their wings, and started rapidly towards it. Other clusters that lay perfectly still, when the first one gave the sound, instantly started from their lethargy, and followed their companions into the hive. Here is *positive* proof, that the sound emitted or produced, was a call to enter the hive, or giving information of one being at hand.

Although out of place, I will here give an omission in the chapter on swarming, which led to my having a swarm of bees lying out all night. It is said, that in extensive bee-gardens in Poland, where many swarms issue at the same time, and preclude the possibility of hiving them separately, that the bees are kept till evening in large boxes, and then emptied out on cloths or sheets in different parcels; and that during the night, the different queens will have collected a cluster around each of them, when the different families may be hived. This appeared so reasonable to me, that I attempted it the last season, for the first time; and a heavy shower came up suddenly, and frustrated my experiment. I had no other opportunity to try it again, but I have no doubt of its being practicable.

Every person that is familiar with bees has, undoubtedly, seen them of a sudden commence the vibration of their wings, standing perfectly still in the mean time. This motion is generally supposed to be an expression of joy, and the only manner in which they can manifest it. I have carefully watched for the cause of this motion of their wings, and my own experience leads me to believe, that the above reason is a correct one. I will give a single proof. Having greatly disturbed a family by turning up the hive, and removing it, by which means, large numbers of bees got astray, flying around in confusion, and on returning it to the stand, the bees immediately flocked around it, and alighting on the floor-board, commenced the vibration of their wings, as above stated; and so continued some minutes. This satisfied me, that it was a sensation of pleasure on again finding their home. This is but one, out of many instances of the same nature, that I have witnessed.

BEES-WAX—HOW MADE.

The nature of wax has already been discussed. My object now is, to show the inexperienced bee-keeper, how to make the article from the combs. The combs are cut out of the hives, the honey secured; and then they are ready for the kettle. Break them in small pieces, or press them into as small a compass as possible, and put them into a woollen bag. Put the bag into the kettle, or vessel of water that is to be set over the fire, and with a flat stone, or some other weight, sink the bag to the bottom. Boil the water about half an hour, then take out the bag, and set the water aside to cool. The wax will rise to the surface. The cake of wax on the surface if containing impurities, may be put into a clean bag, and the second process over the fire, will render it quite clean and pure, and by melting again in some convenient vessel, it may be turned into cups of any shape, first greasing them a little, and when cool, the cakes wil. come out without adhering in the least.

APPENDIX.

MINER'S PATENT EQUILATERAL BEE-HIVE.

IN consequence of the improvement in the ornamental portion of the above hive not being completed, when the original cut was inserted in this work, at page 181, I have concluded to have it appear in an *Appendix*. This is precisely the same hive as that at page 181, except in its embellishments. The size and shape are the same; but it is drawn on a smaller scale, than the other. I consider this the *ne plus ultra* of hives in every point. Nothing of the kind can compare with it in beauty, or in practical value. I do not say this because I am interested; but I say it from a solemn conviction of the truth of the assertion, after having either seen or used almost every other style of hive in existence.

The great value of this hive lies in its internal arrangement. The *nine* communications from the lower to the upper section, are opened and closed at pleasure, *in an instant*, by one of the most simple and valuable inventions imaginable. By the use of this, in connection with other features pertaining to *no other hive,* the management of bees is divested of *every* difficulty. Bees in this hive may be fed, in case of need, with as much ease as a flock of poultry. They must be fed sometimes, when the season has proved unpropitious, but the expense is not as many *shillings* as they will bring in *dollars*, the first good season that occurs.

This hive is intended to occupy any situation that other hives do; either on a shelf or stool. It has a *beveled* bottom-board, thus doing away with the necessity of *suspension*. This kind of bottom-board is of my own invention, as well as *every* part of the hive, and as the right is secured for this, as well as for that represented at page 181, it cannot be constructed except by virtue of a right from me. I have made great improvements in several hives, and which others, perhaps, would also have secured, but I place them before the public in this work, for their *free* use and benefit; but in the hive now in question, I shall claim, and defend my title thereto; even an *imitation* of it *externally*, will not pass with impunity.

Besides the advantages before stated, is that of resting the hive on pinions during summer, and when cold weather arrives, by moving it a quarter of an inch, the whole opening is instantly closed, except a space of two inches in front, and the same in the rear, both of which have perforated slides, so that the bees may be enclosed at pleasure, with a gentle current of air under them. This mode of arrangement is *original* with me, and perhaps I do myself great injustice to give publicity to it, as I have done heretofore in this work; yet I claim it, with the foregoing hive, as a part of my invention, together with the *beveled* bottom-board, and the use of either would be an infringement of my rights; yet in these two points, as valuable as I consider them, I shall not expect the public to be limited in their use of them, so long as my general rights in the Equilateral Hive are not invaded.

This style of hive should be painted white, as that color has much the best appearance on ornamental objects. The chocolate color recommended for other hives, relates to cases where they are merely painted as a protection against the weather.

Here is a pedestal of corresponding architecture, and who will say, that a hive surmounting it, and placed in the flower-garden, would not be a beautiful ornament? If I had to live on a short allowance of food for a year, in order to possess a hive and pedestal of this kind, I would do it, if no other means would obtain them. But let such as have no taste for the elegant and beautiful, have hives of a more common order. This work will suit every taste The *pedestal* does not go with the hive, as a necessary appendage, neither does the *urn*, nor the dental course. The hive may be made perfectly *plain*, if desired, at the cost of ordinary hives, and still possess all its *practical* advantages.

The reader is referred to my advertisement for the price of this hive, &c., in the sequel to this work.

INDEX.

CHAPTER I.

THE QUEEN.

CHAPTER II.

WORKERS.

CHAPTER III.

DRONES.

CHAPTER IV.

EGGS—LARVÆ—TIME TO DEVELOP, ETC.

CHAPTER V.

DIVISION OF LABOR OF BEES.

CHAPTER VI.

BLACK BEES.

CHAPTER VII.

POLLEN, OR BEE-BREAD.

CHAPTER VIII.

WATER AND ITS USES.

CHAPTER IX.

SALT—HOW TO BE USED.

CHAPTER X.

PROPOLIS.

CHAPTER XI.

WAX.

PART SECOND.

CHAPTER XII.

REMARKS.

CHAPTER XIII.

HIVES.

CHAPTER XIV.

BEE-HOUSES.

CHAPTER XV.

BEE-STANDS, ETC.

CHAPTER XVI.

THE APIARY.

CHAPTER XVII.

PASTURAGE.

CHAPTER XVIII.

HONEY DEW.

CHAPTER XIX.

BEE-DRESS, ETC.

CHAPTER XX.

SWARMING, ETC.

CHAPTER XXI.

ARTIFICIAL SWARMS.

CHAPTER XXII.

CHANGING FAMILIES FROM OLD TO NEW HIVES.

CHAPTER XXIII.

THE SEASONS.

CHAPTER XXIV.

PILLAGE OF BEES.

CHAPTER XXV.

GUIDE PLATE, ETC.

CHAPTER XXVI.

MISCELLANEOUS.

APPENDIX.

C. M. SAXTON, PUBLISHER,

121 FULTON STREET, NEW YORK,

WOULD respectfully call attention to his Assortment of Works Appertaining to Agriculture, Rural and Domestic Economy, a few of which are enumerated, with the retail prices, from which a liberal discount will be made when a number of copies are ordered at one time. *Any book* can be sent by mail.

The American Agriculturist, per vol.,	$1.25
Allen's, R. L., American Farm Book,	1.00
Allen's, L. F., American Herd Book,	3.00
Allen's, R. L., Diseases of Domestic Animals,	.75
Allen's, J. F., Treatise on the Grape Vine,	1.00
Hoare on the Vine,	.63
Spooner on the Cultivation of the Grape Vine, and Making of Wine,	.38
Downing's Fruits and Fruit Trees of America,	1.50
Coles' American Fruit Book,	.50
Thomas' Fruit Culturist,	.63
Do. " " with Appendix,	1.00
Ives' New-England Fruit Book,	.50
Bridgman's Fruit Cultivator's Manual,	.50
Iaques' Practical Treatise on the Management of Fruit Trees,	.50
Kenrick's American Orchardist,	.90
Lindley's Guide to the Orchard and Fruit Garden,	1 50
C. M. Hovey's Fruit Trees of America, Colored Plates, per vol.,	6.50
Browne's Trees of America,	4.50
Loudon's Arboretum Britannicum,	55.00
The Complete Gardener and Florist,	.25
Bridgman's Florist's Guide,	.50
Ely's American Florist,	.38
Buist's Flower Garden Directory,	1.25
Sayre's American Flower Garden Companion,	.75
Mrs. Loudon's Companion to the Flower Garden,	1.25
Buist on the Culture of the Rose,	.75
Prince's Rose Manual,	.75
Mrs Gore's Rose Manual,	1.50
Parsons on the Culture of the Rose,	1.50
Rose Culturist,	.38
Lindley's Theory of Horticulture,	1.25
Theodore Thinker's First Lessons in Botany,	.25
Darlington's Agricultural Botany,	1.00
Gray's Botanical Text Book,	1.50
Chapin's Vegetable Kingdom, or Hand Book of Plants,	1.25
Beattie's Essays on Southern Agriculture,	1.00
Woods' Class Book of Botany,	1.50
Partridge's Theory and Practice of Agriculture,	.12½
Rodgers' Scientific Agriculture,	$ .75
Boussingault's Rural Economy,	1.50
Boussingault's Organic Nature,	.50
Falkner's Treatise on the Nature and Value of Manures,	.50
Buel's Farmer's Companion,	75
Buel's Farmer's Instructor, 2 vols.,	1.00
Gaylord and Tucker's American Husbandry,	1.00
Fessenden's Complete Farmer,	.75
Davis' Text Book of Agriculture,	.50
Wiggin's American Farmer's Instructor,	1.50
Pritt's Farmer's Book and Family Instructor,	2.00
Johnson's American Farmer's Encyclopædia,	4.00
Donn's Gardener's Dictionary, 4 vols. quarto,	10.00
Parnell's Applied Chemistry in Arts, Manufactures, and Domestic Economy,	1.00
Fresenius and Bullock's Elementary Instruction in Chemical Analysis,	1.00
Chaptal's Chemistry Applied to Agriculture,	.50
Liebig's Agriculture Chemistry,	.25
Liebig's Animal Chemistry,	.25
Liebig's Familiar Letters on Chemistry,	.25
Topham's Chemistry made Easy for the Agriculturist,	.38
Johnson's Catechism of Agricultural Chemistry and Geology,	.25
Johnson's Lectures on Agricultural Chemistry,	1.25
Skinner's Elements of Agriculture,	.25
Gray's Elements of Scientific and Practical Agriculture,	.50
Robbin's Complete Produce Reckoner, showing the Value, by Pound or Bushel, of all the Different Kinds of Grain,	.75
Whitmarsh on the Mulberry Tree,	.50
Dana's Muck Manual,	.50
Dana's Prize Essay on Manures,	.12½
The Farmer's Mine, or Source of Wealth,	.75
Smith's Productive Farming, or Familiar Digest of Recent Discoveries,	.50
The Farmer's Treasure,	.75
Thompson on the Food of Animals,	.50
The Complete Farrier,	.25
Coles' American Veterinarian,	.50
The American Farrier,	.75
The Horse, its Habits, Diseases, and Management,	.25

Youatt on the Horse, ... $1.75
Miles' Horse's Foot, and How to Keep it Sound, ...25
Hinds' Farrier and Stud Book, . 1.00
Mason's Farrier, ... 1.25
Stewart's Stable Economy, . 1.00
Clater's Every Man His Own Farrier, ...50
Stable Talk and Table Talk, . 1.00
Youatt's Stock Raiser's Manual, 2.50
Clater and Youatt's Cattle Doctor, ...50
Mills' Sportman's Library, . . 1.00
Skinner's Dog and Sportsman, . .75
Hawker and Porter on Shooting, 2.75
Frank Forrister's Field Sports, 4.00
Youatt on the Dog, . . . 1.50
Youatt on the Pig, . . . 65
Knowlson's Cow Doctor, . . .25
Guenon's Treatise on Milch Cows, ...38
Randall's Sheep Husbandry, . 1.00
Morrel's American Shepherd, . 1.00
Canfield on the Management and Breed of Sheep, . . 1.00
Blacklock's Treatise on Sheep, . .50
Bement's American Poulterer's Companion, ... 1.00
Cock's American Poultry Book, .38
Boswell's Poultry Yard, . . .50
Miner's Bee Keeper's Manual, . 1.00
Weeks' Treatise on the Honey Bee, ...50
Bevan on the Bee, ...38
Townley on the Bee, ...50
Marshall's Farmer's and Immigrant's Hand Book, . . 1.00
Stephen's Book of the Farm, 2 vols. octavo, ... 4.00
Ellsworth's Improvements in Arts, Manufactures, &c., in the United States, ...25
Bigelow's Plants of Boston and Vicinity, ... 1.50
Gardiner's Farmer's Dictionary, 1.50
Bement's Journal of Agriculture, ... 2.50
Colman's Continental Agriculture, ... 1.25
Colman's European Agricultural Tour, ... 5.00
Fessenden's New American Gardener, ...84
Mahon's American Gardener's Calendar, ... 3.50
Bridgman's Young Gardener's Assistant, ... 1.75
Johnson's Dictionary of Modern Gardening, ... 2.25
Cobbet's American Gardener, . .38
Bridgman's Kitchen Gardener's Instructor, ...50
Buist's Family Kitchen Gardener, ...75
Thaer's Agriculture, . . . 1.75
Smee on the Potato Plant, . . .72
Gilpin's Landscape Gardening, . 2.50
Downing's Landscape Gardening, ... 3.50
Downing's Cottage Residences, ... 2.00
Lang's Highland Cottages, . . 1.50

Cottage and Villa Architecture, by Walter and Smith, 4 vols. $10.00
Elliot's Cottages & Cottage Life, 2.50
The American Architect, comprising Original Designs of Country Residences, 4to., 1st series, ... 3.50
——— 2d series, ... 3.50
Peters' Agricultural Account Book, ... 1.00
The Canary-Bird Farrier, . . .18¾
Bees, Pigeons, Rabbits, and Canary Birds, ...38
The Bird Keeper's Manual, . .50
The Birds of Long Island, . . 1.00
Gunn's Domestic Medicine, or Poor Man's Friend. This Book points out in plain language, free from Doctor's Terms, the Diseases of Men, Women, and Children, and the Latest and most Approved Means used in their Cure, and is intended expressly for the Benefit of Families. It also Contains a Description of the Medicinal Roots and Herbs in the United States. By John C. Gunn, M. D., 1 vol. 8vo., 3.00
The Use of Brandy and Salt, as a Remedy for Various Internal as well as External Diseases, Inflammation, and Local Injuries, containing Ample Directions for Making and Applying it. By Rev. S. Fenton, ...12½
Miss Beecher's Domestic Economy, ...75
——— Receipt Book, ...75
Miss Leslie's Complete Cookery, 1.20
——— House Book, . . . 1.20
——— Ladies' Receipt Book, . 1.20
——— Indian-Meal Book, . . .25
——— Seventy-Five Receipts, . .30
Mrs. Rundle's Domestic Cookery, ...50
Mrs. Child's Frugal Housewife, .40
The Cook's own Book, . . 1.00
The American Housewife and Kitchen Directory, . . .18¾
The American System of Cookery, ...75
Domestic Cookery, ...50
The Practical Receipt Book, . .62
Miss Acton's New Work on Cookery, ... 1.00
Mrs. Abeel's Skillful Housewife, .25
Mrs. Cornelius' Young Housekeeper's Friend, ...50
Alcott's Young Housekeeper, . 1.00
The Economical Housekeeper, . .75
Browne's Memoir on Indian Corn, .25
Pedder's Farmer's Land Measurer, showing at one View the Contents of any Piece of Land from Dimensions taken in Yards, with a Set of Useful Agricultural Tables, .50
Webster's Encyclopædia of Domestic Economy, . . 3.50

THE AMERICAN BEE-KEEPER'S MANUAL.

BY T. B. MINER.

350 pp. 12mo. 35 ENGRAVINGS. PRICE $1.

PUBLISHED BY C. M. SAXTON, 121 FULTON ST., N. Y.

OPINIONS OF THE PRESS.

"The most complete work on the Bee and Bee-keeping we have yet seen."—*N. Y. Tribune.*

"Mr. Miner has handled this subject in a masterly manner."—*N. Y. True Sun.*

"He has written a work of the most fascinating interest."—*N. Y. Sunday Dispatch.*

"It will interest the general reader. It is indeed a charming volume.—*Commercial Advertiser.*

"This is a truly valuable work, and very interesting."—*Morning Star.*

"It is decidedly the best work we have ever seen."—*Boston Daily Mail.*

"Mr. Miner has performed his task with signal ability."—*Scientific American.*

"It does high credit to the observation and intelligence of the author."—*Christian Intelligencer.*

"This is the most comprehensive and valuable work on the Honey-bee that has ever come under our notice."—*Journal of Commerce.*

"To appreciate the value of the honey-bee one must get this book and read it attentively."—*Noah's Messenger.*

"We like it for its independent tone, and the amount of practical information that it contains."—*Literary World.*

"We have been greatly edified and entertained by this book, from which the reader will collect a great deal of excellent information,—*The Independent.*

"This is probably the most complete manual of the kind ever published. It will richly repay the general reader, too, by the variety of interesting facts it contains."—*Boston Traveller.*

"It is a most excellent and useful treatise, and happily supplies a vacuum that had long existed."—*Boston Times.*

"This volume has all the charm of a romance and admirably displays the habits of this insect."—*Organ.*

"This volume is what it pretends to be, (more than can be said of many works) and all who want a full and thorough history of the nature and management of the bee should have it in their possession."—*Scientific American.*

"It is neatly printed, well illustrated and clearly written and contains a great deal of practical information."—*Mirror.*

"This work probably contains better instructions in regard to bees than any which have ever appeared."—*Sun.*

"The practical directions are the result of evident experience, and being plainly and concisely stated, are excellent, It is so much better than can be obtained elsewhere that we commend it to favor."—*N. Y. Evangelist.*

"It is an excellent book and the best published on the subject."—*Boston Olive Branch.*

THE

AMERICAN POULTRY YARD;

COMPRISING THE

ORIGIN, HISTORY, AND DESCRIPTION OF THE DIFFERENT BREEDS OF

Domestic Poultry;

WITH

Complete Directions for their Breeding, Crossing, Rearing, Fattening, and Preparation for Market; Including Specific Directions for Caponising Fowls, and for the Treatment of the Principal Diseases to which they are subject.

DRAWN FROM AUTHENTIC SOURCES AND PERSONAL OBSERVATION.

Illustrated by Numerous Engravings

BY D. J. BROWNE.

With an Appendix, embracing the Comparative Merits of Different Breed of Fowls.

BY SAMUEL ALLEN.

Price $1, in cloth—75 cts. with paper covers.

OPINIONS OF THE PRESS.

Mr. Browne was bred and brought up a practical farmer. From his intimate knowledge of the history and habits of our domestic animals, having devoted, probably, more attention to the subject, as a whole, by reading and observation, than any other individual in the country, the task of preparing this work was assigned to him.—*Salem Register.*

The style of the engravings and the mechanical execution of the work are excellent. —*Maine Farmer.*

An extensive work on poultry, embracing every information desired.—*N. H. Telegraph.*

We commend the book and the subject to the thoughts of farmers.—*Vt. Watchman.*

Every one who may purchase a copy, upon a perusal of the same, will be fully satisfied that their money was well spent.—*Bristol Phœnix.*

It is one of the best treatises on the Domestic Fowl ever published.—*New-Haven Palladium.*

The details into which this book enters, on all the subjects connected with the profitable raising of fowls, are precisely of that minute and practical character which is needed.—*N. Y. Evangelist.*

We are glad to see that the evident demand for information on the subject of rearing Domestic Fowls has called out so valuable a work as the one before us.—*N. Y. Daily Tribune.*

It contains matter to interest and instruct upon almost everything that concerns the poultry yard, and bears the impress in its pages that its editor was well qualified to fulfill the task he undertook.—*American Farmer.*

It is the most complete book of its class ever published, and quite indispensable to all who are interested in raising poultry.—*Charleston, S. C., Weekly Gazette.*

No farmer among us would remain a day longer without the work, did he but know its value.—*Rahway Register.*

It is unquestionably the cheapest and best work of the kind extant, and should have a place in every farmer's library.—*Germantown, Ohio, Gazette.*

The volume is enriched by an Appendix from the the pen of Mr. Samuel Allen, an experienced breeder of fowls, who has taken much pains to improve the stock in this country.—*N. Y. Weekly Tribune.*

THE AMERICAN BIRD FANCIER;

CONSIDERED WITH REFERENCE TO THE

BREEDING, REARING, FEEDING, MANAGEMENT, AND PECULIARITIES

OF

CAGE AND HOUSE BIRDS.

Illustrated with Engravings.

BY D. J. BROWNE,

AUTHOR OF THE SYLVA AMERICANA, THE AMERICAN POULTRY YARD, ETC.

NEW YORK:

C. M. SAXTON, 121 FULTON STREET.

ALSO, STRINGER & TOWNSEND, H. LONG & BROTHER, W. F. BURGESS,

DEWITT & DAVENPORT, WILSON & CO., DEXTER & BROTHER.

PHILADELPHIA: W. B. ZIEBER, LINDSAY & BLAKISTON.

BOSTON: REDDING & CO.

1850.

www.ingramcontent.com/pod-product-compliance
Lightning Source LLC
LaVergne TN
LVHW010149110826
845151LV00002B/468

* 9 7 8 1 4 2 5 5 3 7 5 4 8 *